投考公務員

能力傾向試

模擬試卷精讀

APTITUDE TEST: MOCK PAPER

精選多份模擬試卷，多達320條題目

投考JRE、CRE、紀律部隊人士適用

Fong Sir 著

序言

公務員薪高糧準，是不少人的理想工作，但無論你是考JRE、CRE，抑或紀律部隊，如要躋身公務員行列，都要成功通過「能力傾向試」(Aptitude Test) 這道必經門檻。

由於公務員事務局一方面沒有「能力傾向試」的試題給考生練習，結果令到考生面對極大壓力。本書的出版，正好為考生提供解決方案。全書多份試卷，合共300多條題目，內容由淺入深，務求幫你反覆操練至最佳狀態，讓你成功在望。

目錄

PART ONE
能力傾向試模擬試卷

CRE-APT

文化會社出版社 **CULTURE CROSS LIMITED**

答題紙 ANSWER SHEET

請在此貼上電腦條碼
Please stick the barcode label here

(1) 考生編號 Candidate No.

(2) 考生姓名 Name of Candidate

(3) 考生簽署 Signature of Candidate

宜用H.B.鉛筆作答
You are advised to use H.B. Pencils

考生須依照下圖所示填畫答案：

23 A B C D E

錯填答案可使用潔淨膠擦將筆痕徹底擦去。

切勿摺皺此答題紙

Mark your answer as follows:

23 A B C D E

Wrong marks should be completely erased with a clean rubber.

DO NOT FOLD THIS SHEET

No.					
1	A	B	C	D	E
2	A	B	C	D	E
3	A	B	C	D	E
4	A	B	C	D	E
5	A	B	C	D	E
6	A	B	C	D	E
7	A	B	C	D	E
8	A	B	C	D	E
9	A	B	C	D	E
10	A	B	C	D	E
11	A	B	C	D	E
12	A	B	C	D	E
13	A	B	C	D	E
14	A	B	C	D	E
15	A	B	C	D	E
16	A	B	C	D	E
17	A	B	C	D	E
18	A	B	C	D	E
19	A	B	C	D	E
20	A	B	C	D	E
21	A	B	C	D	E
22	A	B	C	D	E
23	A	B	C	D	E
24	A	B	C	D	E
25	A	B	C	D	E
26	A	B	C	D	E
27	A	B	C	D	E
28	A	B	C	D	E
29	A	B	C	D	E
30	A	B	C	D	E
31	A	B	C	D	E
32	A	B	C	D	E
33	A	B	C	D	E
34	A	B	C	D	E
35	A	B	C	D	E
36	A	B	C	D	E
37	A	B	C	D	E
38	A	B	C	D	E
39	A	B	C	D	E
40	A	B	C	D	E

文 化 會 社 出 版 社
投 考 公 務 員　模 擬 試 題 王

能力傾向試測試
模擬試卷（一）

時間：四十五分鐘

考生須知：

（一）細讀答題紙上的指示。宣布開考後，考生須首先於適當位置貼上電腦條碼及填上各項所需資料。宣布停筆後，考生不會獲得額外時間貼上電腦條碼。

（二）試場主任宣布開卷後，考生請檢查試題冊及確定試題冊內的試題。最後會有「**全卷完**」的字眼。

（三）本試卷各題佔分相等。

（四）**本試卷全部試題均須回答**。為便於修正答案，考生宜用HB鉛筆把答案填畫在答題紙上。錯誤答案可用潔淨膠擦將筆痕徹底擦去。考生須清楚填畫答案，否則會因答案未能被辨認而失分。

（五）每題只可填畫**一個**答案。如填劃超過一個答案，該題將**不獲評分**。

（六）答案錯誤，不另扣分。

（七）未經許多，請勿打開試題冊。

CC-CRE-APT

I. 演繹推理（8題）

請根據以下短文的內容，選出一個或一組推論。請假定短文的內容都是正確的。

1. 1998年9月，德國社會民主黨人施羅德擊敗連續執政16年之久的科爾，出任德國總理。由此，歐盟迎來了歷史上第一次德、法、英三個歐洲大國都由左派執掌政府的時代。現在歐盟15國除西班牙和愛爾蘭外均由左派單獨執政或左派聯合中、右派共同執政，其中11國的政府首腦是左派人士。

   據此，我們可以知道：

   A. 在本世紀即將結束之際，世界各國的左翼政黨全面得勢。

   B. 在歐盟中，至少有11個國家為左派單獨執政。

   C. 在歐盟中，右派單獨或參與執政的國家不超過11個。

   D. 在歐盟中，至少有3個國家的右派參加執政。

2. 甲、乙、丙都是體育愛好者。夏天，三人分別參加了A城的羽毛球、B城的游泳和C城的划艇三項體育比賽，並都得了冠軍。已知，甲沒有到C城划艇，丙沒參加B城的游泳比賽，划艇冠軍不是丙。據此，我們可以知道：

   A. 甲是游泳冠軍

   B. 乙是游泳冠軍

   C. 乙沒去C城划艇

   D. 丙不是羽毛球冠軍

3. **有一份選擇題試卷共有6條題目，其得分標準是：一條小題答對得8分，答錯0分，不答(沒填任何答案)2分，某位同學得了20分，則他：**

A. 至多答對一條題

B. 至少有三條小題沒答

C. 至少答對三條小題

D. 答錯兩條小題

4. **某保險公司推出一項「希望之星」保險計劃，該業務要求5歲兒童的家長每年交600元保費並至14歲，即可享受兒童將來讀大學的所有學費，則促使家長不參加投保的最適當理由為：**

A. 家長並不能確定兒童將來上哪一所大學

B. 10年的累計保險費總額大於將來上大學的學費總和

C. 據估計，每年上大學的學費增長率大於生活費的增長率

D. 該業務並不包括支付上大學的食宿費用

5. 對建築和製造業的安全研究表明，企業工作負荷量加大時，工傷率也隨之提高。工作負荷增大時，企業總是僱用大量的不熟練的工人，毫無疑問，工傷率上升是由非熟練工的低效率造成的。則能夠對上述觀點作出最強的反駁的一項是：

A. 負荷量增加時，企業僱用非熟練工只是從事臨時性的工作

B. 建築業的工傷率比製造業高

C. 只要企業工作負荷增加，熟練工人的事故率總是隨之增高

D. 需要僱用新人的企業應加強職業訓練

6. 令人奇怪的是，洛可可風格竟然首先出現於法蘭西。路易十四的統治持續時間太長，對老王朝過分虔誠的時代終於結束，雄偉高貴的凡爾賽不再迫使人們參加令人生厭的慶典，從此人們聚集於巴黎各公館的精美沙龍之中。起初，洛可可是一種新型裝飾，是為熱愛冒險、異國情調、奇思遐想和大自然的高雅、智慧的社會服務的。這種輕盈、精美的風格最適合於現代公寓，房間不大，但均有珍貴的用途。洛可可進入巴洛克風格出盡風頭的國家時，是十分謹慎小心的。
從這段話可以得到的正確推論是：

A. 路易十四時期法國流行輕盈、精美的巴洛克式室內裝潢風格

B. 洛可可風格首先在法國出現是因為法國人熱愛冒險和奇思遐想

C. 現代公寓都是洛可可式的，因而它們房間不大，但均有珍貴的用途

D. 與巴洛克風格相比，洛可可風格並不追求雄偉和高貴

7. 體育館內正進行一場乒乓球雙打比賽，觀眾議論雙方運動員甲、乙、丙、丁的年齡：

(1)「乙比甲的年齡大」

(2)「甲比他的伙伴的年齡大」

(3)「丙比他的兩個對手的年齡都大」

(4)「甲與乙的年齡差距比丙與乙的年齡差距更大些」

根據這些議論，甲、乙、丙、丁的年齡從大到小的順序是：

A. 甲、丙、乙、丁

B. 丙、乙、甲、丁

C. 乙、甲、丁、丙

D. 乙、丙、甲、丁

8. 某流動通訊公司曾經投入巨資擴大流動通訊服務覆蓋率，結果當年用戶增加了25%，但是總利潤卻下降了10%。最可能的原因是：

A. 該流動通訊公司新增用戶的消費總額相對較低

B. 該流動通訊公司的電話月費大幅度下降了

C. 該流動通訊公司當年的管理出了問題

D. 該流動通訊公司為擴大市場佔有率，投入過多資金

II. Verbal Reasoning (English) (6 questions)

Directions :
In this test, each passage is followed by three statements (the questions). You have to assume what is stated in the passage is true and decide whether the statements are either:
True (Box A): the statement is already made or implied in the passage, or follows logically from the passage.
False (Box B): the statement contradicts what is said, implied by, or follows logically from the passage.
Can’ t tell (Box C): there is insufficient information in the passage to establish whether the statement is true or false.

Passage 1 (Question 9 to 11):

Instituted in 1979 as a temporary measure to limit population growth, China's one child policy remains in force today and is likely to continue for another decade.
China's population control policy has attracted criticism because of the manner in which it is enforced, and also because of its social repercussions. Supporters of the Chinese government's policy consider it a necessary measure to curb extreme overpopulation, which has resulted in a reduction of an estimated 300 million people in its first twenty years. Not only is a reduced population environmentally beneficial, it also increases China's per capita gross domestic product. The one-child policy has led to a disparate ratio of males to females – with abortion, abandonment and infanticide

of female infants resulting from a cultural preference for sons. Furthermore, Draconian measures such as forced sterilization are strongly opposed by critics as a violation of human reproduction rights. The one-child policy is enforced strictly in urban areas, whereas in provincial regions fines are imposed on families with more than one child. There are also exceptions to the rules – for example, ethnic minorities. A rule also allows couples without siblings to have two children – a provision which applies to millions of sibling-free adults now of child-bearing age.

9. China's one-child policy increases the country's wealth.

10. The passage suggests that two-child families will dramatically increase, as sibling-free adults reach child-bearing age.

11. The main criticism of China's one-child policy is that it violates human rights.

Passage 2 (Question 12 to 14):

Brand equity has become a key asset in the world of competitive business. Indeed, some brands are now worth more than companies. Large corporations themselves are widely distrusted, whereas strangely, brands have the opposite effect on people. Brands are used to humanise corporations by appropriating characteristics such as courage, honesty, friendliness and fun. An example is Dove soap, where a dove represents white, cleanliness and peace. Volkswagen like to give the impression through their advertising that they are a reliable, clever, technical product. In a sense, rather than the product itself, the image and the idea are the selling point.

12. Brands have always been an important asset to a company.

13. Many people distrust large corporations.

14. Dove soap chose a dove for their brand to give a sense of cleanliness and peace.

III. Data Sufficiency Test (8 questions)

Directions : In this test, you are required to choose a combination of clues to solve a problem.

15. What is Monica’ s position with respect to Rahul?

    (1) In a row of 25 students, Monica is sitting 12th from right end of row and Rahul is sitting 20th from left end of the row

    (2) Monica is 4th from right end and Rahul is 8th from left end

    A. If data in the statement (1) alone is sufficient to answer the question

B. If data in the statement (2) alone is sufficient to answer the question

C. If data either in the statement (1) alone or statement (2) alone are sufficient to answer the question

D. If data given in both (1) & (2) together are not sufficient to answer the question

E. If data in both statements (1) & (2) together are necessary to answer the question

16. Rohit, Kajol, Tanmay and Suman are four friends. Who is the oldest among them?

(1) The total age of Kajol and Tanmay together is more than that of Suman

(2) The total age of Rohit and Kajol together is less than that of Suman

A. if the data in statement (1) alone are sufficient to answer the question

B. if the data in statement (2) alone are sufficient answer the question

C. if the data either in (1) or (2) alone are sufficient to answer the question

D. if the data even in both the statements together are not sufficient to answer the question

E. If the data in both the statements together are needed

17. How many visitors saw the exhibition yesterday?

(1) Each entry pass holder can take up to three persons with him/her

(2) In all, 243 passes were sold yesterday

A. if the data in statement (1) alone are sufficient to answer the question

B. if the data in statement (2) alone are sufficient answer the question

C. if the data either in (1) or (2) alone are sufficient to answer the question

D. if the data even in both the statements together are not sufficient to answer the question

E. If the data in both the statements together are needed

18. Hemanth ranks tenth in a class. How many students are there in the class?

(1) His friend got 58th rank which is the last

(2) Hemanth's rank from the last is 49th

A. if the data in statement (1) alone are sufficient to answer the question

B. if the data in statement (2) alone are sufficient answer the question

C. if the data either in (1) or (2) alone are sufficient to answer the question

D. if the data even in both the statements together are not sufficient to answer the question

E. If the data in both the statements together are needed

19. What is Gagan's age?

(1) Gagan, Vimal and Kunal are all of the same age

(2) Total age of Vimal, Kunal and Anil is 32 and Anil is as old as Vimal and Kunal together

A. if the data in statement I alone are sufficient to answer the question

B. if the data in statement II alone are sufficient answer the question

C. if the data either in I or II alone are sufficient to answer the question

D. if the data even in both the statements together are not sufficient to answer the question

E. If the data in both the statements together are needed

20. Who is C's partner in a game of cards involving four players A, B, C and D?

(1) D is sitting opposite to A.

(2) B is sitting right of A and left of D

A. if the data in statement (1) alone are sufficient to answer the question

B. if the data in statement (2) alone are sufficient answer the question

C. if the data either in (1) or (2) alone are sufficient to answer the question

D. if the data even in both the statements together are not sufficient to answerthe question

E. If the data in both the statements together are needed

21. Total money with Naresh and Ajay is 28 percent of that with Usman. How much money is Ajay having?

(1) Usman has got Rs.75000

(2) The ratio of money of Naresh to money held by Ajay is 1:3

A. if the data in statement (1) alone are sufficient to answer the question

B. if the data in statement (2) alone are sufficient answer the question

C. if the data either in (1) or (2) alone are sufficient to answer the question

D. if the data even in both the statements together are not sufficient to answer the question

E. If the data in both the statements together are needed

22. How many gift boxes were sold on Monday?

(1) It was 10% more than the boxes sold on the earlier day i.e., Sunday

(2) Every third visitor to the shop purchased the box and 1500 visitors were there on Sunday

A. if the data in statement (1) alone are sufficient to answer the question

B. if the data in statement (2) alone are sufficient answer the question

C. if the data either in (1) or (2) alone are sufficient to answer the question

D. if the data even in both the statements together are not sufficient to answer the question

E. If the data in both the statements together are needed

IV. Numerical Reasoning (5 questions)

Directions :

Each question is a sequence of numbers with one or two numbers missing. You have to figure out the logical order of the sequence to find out the missing number(s).

23. 0，3，2，5，4，7，( )

A. 6

B. 7

C. 8

D. 9

24. 7，14，5，15，3，12，2，( )

A. 4

B. 10

C. 5

D. 6

25. 1，2，3，7，16，( )

A. 66

B. 65

C. 64

D. 63

26. 0，1，3，8，22，63，( )

A. 122

B. 174

C. 185

D. 196

27. 5，3，4，1，9，( )

A. 24

B. 11

C. 37

D. 64

V. Interpretation of Tables and Graphs (8 questions)

Directions :

This is a test on reading and interpretation of data presented in tables and graphs.

Chart 1

Study the following graph carefully and answer the questions given below:

Distribution of candidates who were enrolled for MBA entrance exam and the candidates (out of those enrolled. who passed the exam in different institutes:

Candidates Enrolled = 8550

X 16%
P 22%
V 12%
Q 15%
T 8%
S 17%
R 10%

Candidates who Passed the Exam = 5700

X 12%
P 18%
V 15%
Q 17%
T 9%
S 16%
R 13%

28. What percentage of candidates passed the Exam from institute T out of the total number of candidates enrolled from the same institute?

A. 50%

B. 62.5%

C. 75%

D. 80%

29. Which institute has the highest percentage of candidates passed to the candidates enrolled?

A. Q

B. R

C. V

D. T

30. The number of candidates passed from institutes S and P together exceeds the number of candidates enrolled from institutes T and R together by:

A. 228

B. 279

C. 399

D. 407

31. What is the percentage of candidates passed to the candidates enrolled for institutes Q and R together?

A. 68%

B. 80%

C. 74%

D. 65%

Chart 2

Two different finance companies declare fixed annual rate of interest on the amounts invested with them by investors. The rate of interest offered by these companies may differ from year to year depending on the variation in the economy of the country and the banks rate of interest. The annual rate of interest offered by the two Companies P and Q over the years are shown by the line graph provided below.

Annual Rate of Interest Offered by Two Finance Companies Over the Years.

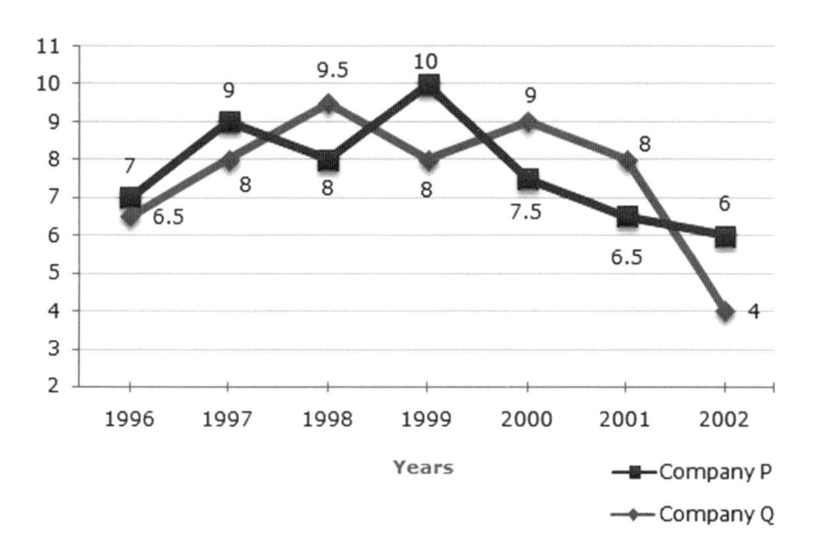

32. A sum of Rs. 4.75 lakhs was invested in Company Q in 1999 for one year. How much more interest would have been earned if the sum was invested in Company P?

A. Rs. 19,000

B. Rs. 14,250

C. Rs. 11,750

D. Rs. 9500

33. If two different amounts in the ratio 8:9 are invested in Companies P and Q respectively in 2002, then the amounts received after one year as interests from Companies P and Q are respectively in the ratio?

A. 2:3

B. 3:4

C. 6:7

D. 4:3

34. In 2000, a part of Rs. 30 lakhs was invested in Company P and the rest was invested in Company Q for one year. The total interest received was Rs. 2.43 lakhs. What was the amount invested in Company P?

A. Rs. 9 lakhs

B. Rs. 11 lakhs

C. Rs. 12 lakhs

D. Rs. 18 lakhs

35. An investor invested a sum of Rs. 12 lakhs in Company P in 1998. The total amount received after one year was re-invested in the same Company for one more year. The total appreciation received by the investor on his investment was?

A. Rs. 2,96,200

B. Rs. 2,42,200

C. Rs. 2,25,600

D. Rs. 2,16,000

全卷完

CRE-APT

文化會社出版社 **CULTURE CROSS LIMITED**

答題紙 ANSWER SHEET

請在此貼上電腦條碼
Please stick the barcode label here

(1) 考生編號 Candidate No.

(2) 考生姓名 Name of Candidate

(3) 考生簽署 Signature of Candidate

宜用H.B. 鉛筆作答
You are advised to use H.B. Pencils

考生須依照下圖所示填畫答案：

23 A B C D E

錯填答案可使用潔淨膠擦將筆痕徹底擦去。

切勿摺皺此答題紙

Mark your answer as follows:

23 A B C D E

Wrong marks should be completely erased with a clean rubber.

DO NOT FOLD THIS SHEET

1	A	B	C	D	E	21	A	B	C	D	E
2	A	B	C	D	E	22	A	B	C	D	E
3	A	B	C	D	E	23	A	B	C	D	E
4	A	B	C	D	E	24	A	B	C	D	E
5	A	B	C	D	E	25	A	B	C	D	E
6	A	B	C	D	E	26	A	B	C	D	E
7	A	B	C	D	E	27	A	B	C	D	E
8	A	B	C	D	E	28	A	B	C	D	E
9	A	B	C	D	E	29	A	B	C	D	E
10	A	B	C	D	E	30	A	B	C	D	E
11	A	B	C	D	E	31	A	B	C	D	E
12	A	B	C	D	E	32	A	B	C	D	E
13	A	B	C	D	E	33	A	B	C	D	E
14	A	B	C	D	E	34	A	B	C	D	E
15	A	B	C	D	E	35	A	B	C	D	E
16	A	B	C	D	E	36	A	B	C	D	E
17	A	B	C	D	E	37	A	B	C	D	E
18	A	B	C	D	E	38	A	B	C	D	E
19	A	B	C	D	E	39	A	B	C	D	E
20	A	B	C	D	E	40	A	B	C	D	E

文化會社出版社
投考公務員 模擬試題王

能力傾向試測試
模擬試卷（二）

時間：四十五分鐘

考生須知：

（一）細讀答題紙上的指示。宣布開考後，考生須首先於適當位置貼上電腦條碼及填上各項所需資料。宣布停筆後，考生不會獲得額外時間貼上電腦條碼。

（二）試場主任宣布開卷後，考生請檢查試題冊及確定試題冊內的試題。最後會有「**全卷完**」的字眼。

（三）本試卷各題佔分相等。

（四）**本試卷全部試題均須回答**。為便於修正答案，考生宜用HB鉛筆把答案填畫在答題紙上。錯誤答案可用潔淨膠擦將筆痕徹底擦去。考生須清楚填畫答案，否則會因答案未能被辨認而失分。

（五）每題只可填畫**一個**答案。如填劃超過一個答案，該題將**不獲評分**。

（六）答案錯誤，不另扣分。

（七）未經許多，請勿打開試題冊。

I. 演繹推理（8題）

請根據以下短文的內容，選出一個或一組推論。請假定短文的內容都是正確的。

1. **發達國家中冠心病的發病率大約是發展中國家的三倍。有人認為，這主要歸咎於發達國家中人們的高脂肪、高蛋白、高熱量的食物攝取。相對來說，發展中國家較少有人具備生這種「富貴病」的條件。其實，這種看法很難成立。因為，目前發達國家的人均壽命高於70歲，而發展中國家的人均壽命還不到50歲。以下哪項如果成立，最能加強上述反駁？**

A. 統計資料顯示，冠心病患者相對集中在中老年的年齡階段，即45歲以上

B. 目前冠心病患者呈年輕化趨勢

C. 發展中國家人們的高脂肪、高蛋白、高熱量食物的攝入量，無論是總量還是人均量，都在逐年增長

D. 相對發展中國家來說，發達國家的人們具有較高的防治冠心病的常識和較好的醫療條件

2. **某國的經濟以農業為主，2018年遭受百年不遇的旱災，國際有關組織號召各國人民向該國伸出援助之手。下面最可能的推斷是：**

A. 該國農業生產力水平低

B. 該國是一個經濟發達國家

C. 該國國民收入低，生活水平低

D. 該國的經濟受氣候影響大

3. **如果某人是殺人犯，那麼案發時他在現場。據此，我們可以推出：**

A · 張先生案發時在現場，所以他是殺人犯

B · 李先生不是殺人犯，所以他案發時不在現場

C · 王先生案發時不在現場，所以他不是殺人犯

D · 許先生不在案發現場，但他是殺人犯

4. **在農業發展初期，很少遇到昆蟲問題。這一問題是隨著農業的發展而產生的——在大面積土地上僅種一種穀物，這樣的種植方法為某些昆蟲的猛增提供了有利條件。很明顯，一種食麥昆蟲在專種麥子的農田裡，比在其他農田裡繁殖起來要快得多。上述論斷不能解釋下列哪種情況？**

A. 一種由甲蟲帶來的疾病掃蕩了某城市街道兩旁的梧桐樹

B. 控制某一種類生物的棲息地的適宜面積，符合自然發展規律的格局

C. 遷移到新地區的物種，由於逃離了其天敵對它的控制而蓬勃發展起來

D. 楊樹的害蟲在與其他樹木摻雜混種的楊樹林中的繁殖速度，會受到限制

5. 與新疆的其他城市一樣，庫爾勒直至20世紀80年代初物價都是很低。自它成為新疆的石油開採中心後，那裡的物價大幅上升。這種物價上漲可能來自這場石油經濟，這是因為新疆那些沒有石油經濟的城市，仍然保持著很低的物價水平。

最準確的描述了上段論述中所採用的推理方法的一項是：

A. 鑒於條件不存在的時候現象沒有發生，所以認為條件是現象的一個原因

B. 鑒於有時條件不存在的情況下現象也會發生，所以認為條件不是現象的前提

C. 由於某一特定事件在現象發生前沒有出現，所以認為這一事件不可能引發現象

D. 試圖説明某種現象是不可能發生的，而某種解釋正確就必須要求這種現象發生

6. 中國是農業國家，農業是國民經濟基礎。減輕農民負擔，就是要保護和調動農民積極性，促進農業、農村經濟和國民經濟的發展。如果不注意保護農民利益，隨意向農民亂收費、亂罰款和進行各類集資攤派，必將挫傷農民生產積極性。

所以：

A. 要發展經濟，特別是發展農村基礎設施，就要增加農民負擔

B. 發展經濟與減輕農民負擔兩者並不矛盾

C. 不減輕農民負擔，將會影響農村的社會穩定

D. 農民負擔問題是農村經濟發展的最主要問題

7. **經警方對甲、乙、丙和丁四人錄取口供，已知下列判斷為真：**

(1) 若甲和乙都是殺人犯，則丙是無罪

(2) 丙有罪，並且丁的陳述正確

(3) 只有丁的陳述不正確，乙才不是殺人犯

由此可以推出下列哪項是正確的：

A. 甲、丙是殺人犯

B. 丙、丁是殺人犯

C. 甲不是殺人犯，乙是殺人犯

D. 甲是殺人犯，乙不是殺人犯

8. **一位雄心勃勃的年輕人想發明一種能夠溶解一切物質的溶液。下面哪項勸告最能使這位年輕人改變初衷呢？**

A. 許多人都已經對此做過嘗試，沒有一個是成功的

B. 理論研究證明這樣一種溶液是不存在的

C. 研究此溶液需要複雜的工藝和設備，你的條件不具備

D. 這種溶液研製出來以後，你打算用什麼容器來盛放它呢

II. Verbal Reasoning (English) (6 questions)

Directions :
In this test, each passage is followed by three statements (the questions). You have to assume what is stated in the passage is true and decide whether the statements are either:
True (Box A): the statement is already made or implied in the passage, or follows logically from the passage.
False (Box B): the statement contradicts what is said, implied by, or follows logically from the passage.
Can't tell (Box C): there is insufficient information in the passage to establish whether the statement is true or false.

Passage 1 (Question 9 to 11):

Sodium chloride, or salt, is essential for human life. Typically derived from the evaporation of sea water or the mining of rock salt deposits, salt has been used by humans for thousands of years as a food seasoning and preservative. The mineral sodium is an electrolyte – an electrically-charged ion – that enables cells to carry electrical impulses to other cells, for example muscle contractions. Electrolytes also regulate the body's fluid levels. A diet deficient in salt can cause muscle cramps, neurological problems and even death. Conversely, a diet high in salt leads to an increased risk of conditions such as hypertension, heart disease and stroke. In spite of high-profile campaigns to raise awareness, salt consumption has increased by 50% in the past four decades, with the average adult ingesting more than double the amount of salt their body requires.

Much of this increase can be attributed to the advent of frozen and processed foods in the mid-twentieth century. In the United States it is estimated that excessive salt consumption claims 150,000 lives and results in $24 billion of health care costs annually. For individuals wishing to reduce their sodium intake, the answer is not simply rejecting the salt shaker; 75% of the average person's salt consumption comes from food, such as bread, cereals, and cheese.

9. Humans primarily use salt for food flavouring and preservation.

10. Most adults consume 50% more salt than their body requires.

11. Frozen and processed foods contain no more salt than contained in a typical diet.

Passage 2 (Question 12 to 14):

The costs of roaming - the service which allows UK customers to use their mobile phone abroad - are much higher than those in France, Germany, Sweden and Italy. Many people get caught out because they are unaware of the high prices, and that they get charged for simply receiving calls whilst abroad. Less than a quarter of consumers had any knowledge of the price of using a mobile phone aboard when they bought their phone. Better consumer information is vital if prices for pre-pay international roaming in the UK are to come down.

12. It is more expensive for German customers to use a roaming service than it is for UK customers.

13. Pay-as-you-go roaming rates are lower than they are for contract customers.

14. Customers can be charged for calls which they do not make themselves.

III. Data Sufficiency Test (8 questions)

Directions : In this test, you are required to choose a combination of clues to solve a problem.

15. How many new year' s greeting cards were sold this year in your shop?

    (1) Last year 2935 cards were sold

    (2) The number of cards sold this year was 1.2 times that of last year

    A. if the data in statement (1) alone are sufficient to answer the question

B. if the data in statement (2) alone are sufficient answer the question

C. if the data either in (1) or (2) alone are sufficient to answer the question

D. if the data even in both the statements together are not sufficient to answer the question

E. If the data in both the statements together are needed

16. How many sons does D have?

(1) A’ s father has three children

(2) B is A’ s brother and son of D

A. if the data in statement (1) alone are sufficient to answer the question

B. if the data in statement (2) alone are sufficient answer the question

C. if the data either in (1) or (2) alone are sufficient to answer the question

D. if the data even in both the statements together are not sufficient to answer the question

E. If the data in both the statements together are needed

17. What is the code for 'or' in the code language?

(1) 'nik sa te' means 'right or wrong', 'ro da nik' means 'he is right' and 'fe te ro' means 'that is wrong'.

(2) 'pa nik la' means 'that right man', 'sa ne pa' means 'this or that' and 'ne ka re' means 'tell this there'.

A. If the data in statement (1) alone are sufficient

B. If the data in statement (2) alone are sufficient

C. If the data either in statement (1) alone or in statement (2) alone are sufficient

D. If the data given in both the statements (1) and (2) together are not sufficient

E. If the data in both the statements (1) and (2) together are necessary

18. A shopkeeper sells a phone X on 20% profit. What is the Cost Price of phone X?

(1) Average of all the phones sold by shopkeeper is 225

(2) Selling Price of Phone X is Rs. 14400

A. if statement (1) alone is sufficient but B alone is not sufficient

B. if statement (2) alone is sufficient but statement (1) alone is not sufficient

C. if both the statements (1) and (2) are needed

D. if each of statement (1) or statement (2) alone is sufficient

E. if both of them together are not sufficient

19. What could be the maximum number of people in a group reading both newspapers X & Y?

(1) 50% of persons of the group read newspaper X

(2) 30% of persons of the group read newspaper Y

A. if statement (1) alone is sufficient but B alone is not sufficient

B. if statement (2) alone is sufficient but statement (1) alone is not sufficient

C. if both the statements (1) and (2) are needed

D. if each of statement (1) or statement (2) alone is sufficient

E. if both of them together are not sufficient

20. In which month is Meena's birthday?

(1) Shikha remembers that Meena's birthday was 4 months ago

(2) Raj remembers that after 2 months from now, Meena's birthday will be 6 months back

A. If data in the statement (1) alone is sufficient to answer the question

B. If data in the statement (2) alone is sufficient to answer the question

C. If data either in the statement (1) alone or statement (2) alone are sufficient to answer the question

D. If data given in both (1) & (2) together are not sufficient to answer the question

E. If data in both statements (1) & (2) together are necessary to answer the question

21. Amit is facing which direction?

(1) Shikha is facing east direction and if she turns to her right she will face Raj

(2) Amit is facing opposite direction as that of Kiran who is facing Shikha

A. If data in the statement (1) alone is sufficient to answer the question

B. If data in the statement (2) alone is sufficient to answer the question

C. If data either in the statement (1) alone or statement (2) alone are sufficient to answer the question

D. If data given in both (1) & (2) together are not sufficient to answer the question

E. If data in both statements (1) & (2) together are necessary to answer the question

22. Question 8. On which floor is Shikha residing?

(1) In a six storey building (Ground floor is parking space), Rekha is on fourth floor. Shikha likes to reside only on even numbered floors. Reema is not on the topmost floor.

(2) Reema is two floors below Peter who is 3 floors abovc Shikha.

A. If data in the statement (1) alone is sufficient to answer the question

B. If data in the statement (2) alone is sufficient to answer the question

C. If data either in the statement (1) alone or statement (2) alone are sufficient to answer the question

D. If data given in both (1) & (2) together are not sufficient to answer the question

E. If data in both statements (1) & (2) together are necessary to answer the question

IV. Numerical Reasoning (5 questions)

Directions :

Each question is a sequence of numbers with one or two numbers missing. You have to figure out the logical order of the sequence to find out the missing number(s).

23. 1，3，2，3，4，9，( )

A. 13

B. 18

C. 29

D. 32

24. -8，15，39，65，94，128，170，(　)

A. 180

B. 210

C. 225

D. 256

25. 1，1，5，7，13，(　)

A. 15

B. 17

C. 19

D. 21

26. 5/6，6/11，11/17，(　)，28/45

A. 17/24

B. 17/28

C. 28/17

D. 24/17

27. 3，4，6，12，36，(　)

A. 72

B. 108

C. 216

D. 288

V. Interpretation of Tables and Graphs (8 questions)

Directions :
This is a test on reading and interpretation of data presented in tables and graphs.

The following table gives the sales of batteries manufactured by a company over the years.

Chart 1

Number of Different Types of Batteries Sold by a Company Over the Years (Numbers in Thousands)

Year	Types of Batteries					
	4AH	7AH	32AH	35AH	55AH	Total
1992	75	144	114	102	108	543
1993	90	126	102	84	126	528
1994	96	114	75	105	135	525
1995	105	90	150	90	75	510
1996	90	75	135	75	90	465
1997	105	60	165	45	120	495
1998	115	85	160	100	145	605

28. What was the approximate percentage increase in the sales of 55AH batteries in 1998 compared to that in 1992?

    A. 28%

    B. 31%

    C. 33%

    D. 34%

29. The total sales of all the seven years is the maximum for which battery?

A. 4AH

B. 7AH

C. 32AH

D. 35AH

30. What is the difference in the number of 35AH batteries sold in 1993 and 1997?

A. 24000

B. 28000

C. 35000

D. 39000

31. The percentage of 4AH batteries sold to the total number of batteries sold was maximum in the year?

A. 1994

B. 1995

C. 1996

D. 1997

Chart 2

The bar graph given below shows the percentage distribution of the total production of a car manufacturing company into various models over two years.

Percentage of Six different types of Cars manufactured by a Company over Two Years

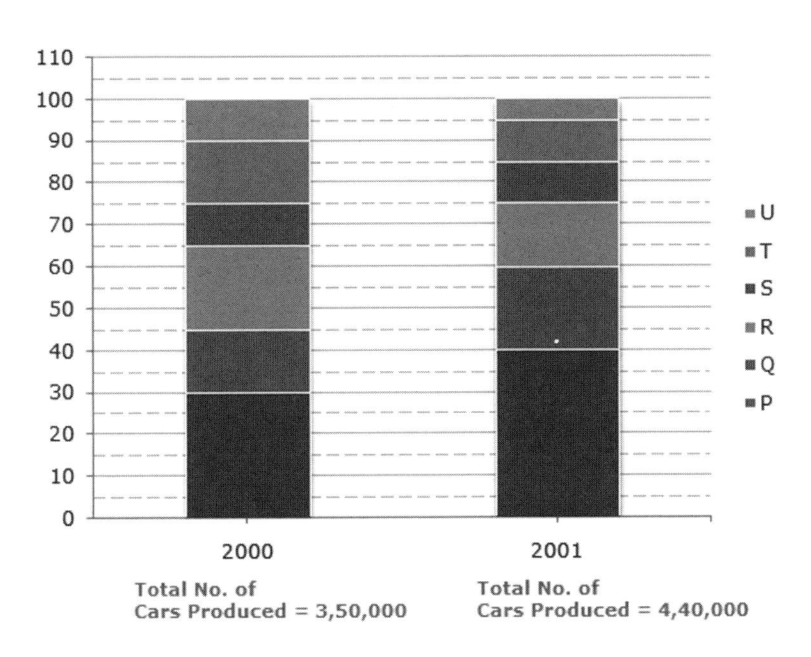

32. What was the difference in the number of Q type cars produced in 2000 and that produced in 2001?

A. 35,500

B. 27,000

C. 22,500

D. 17,500

33. Total number of cars of models P, Q and T manufactured in 2000 is?

A. 245,000

B. 227,500

C. 210,000

D. 192,500

34. If the percentage production of P type cars in 2001 was the same as that in 2000, then the number of P type cars produced in 2001 would have been?

A. 140,000

B. 132,000

C. 117,000

D. 105,000

35. If 85% of the S type cars produced in each year were sold by the company, how many S type cars remain unsold?

A. 7650

B. 9350

C. 11,850

D. 12,250

全卷完

CRE-APT

文化會社出版社 **CULTURE CROSS LIMITED**

答題紙 ANSWER SHEET

請在此貼上電腦條碼 Please stick the barcode label here

(1) 考生編號 Candidate No.

(2) 考生姓名 Name of Candidate

(3) 考生簽署 Signature of Candidate

宜用H.B.鉛筆作答
You are advised to use H.B. Pencils

考生須依照下圖所示填畫答案：

23 A B C D E

錯填答案可使用潔淨膠擦將筆痕徹底擦去。

切勿摺皺此答題紙

Mark your answer as follows:

23 A B C D E

Wrong marks should be completely erased with a clean rubber.

DO NOT FOLD THIS SHEET

1	A	B	C	D	E	21	A	B	C	D	E
2	A	B	C	D	E	22	A	B	C	D	E
3	A	B	C	D	E	23	A	B	C	D	E
4	A	B	C	D	E	24	A	B	C	D	E
5	A	B	C	D	E	25	A	B	C	D	E
6	A	B	C	D	E	26	A	B	C	D	E
7	A	B	C	D	E	27	A	B	C	D	E
8	A	B	C	D	E	28	A	B	C	D	E
9	A	B	C	D	E	29	A	B	C	D	E
10	A	B	C	D	E	30	A	B	C	D	E
11	A	B	C	D	E	31	A	B	C	D	E
12	A	B	C	D	E	32	A	B	C	D	E
13	A	B	C	D	E	33	A	B	C	D	E
14	A	B	C	D	E	34	A	B	C	D	E
15	A	B	C	D	E	35	A	B	C	D	E
16	A	B	C	D	E	36	A	B	C	D	E
17	A	B	C	D	E	37	A	B	C	D	E
18	A	B	C	D	E	38	A	B	C	D	E
19	A	B	C	D	E	39	A	B	C	D	E
20	A	B	C	D	E	40	A	B	C	D	E

文化會社出版社
投考公務員 模擬試題王

能力傾向試測試
模擬試卷（三）

時間：四十五分鐘

考生須知：

(一) 細讀答題紙上的指示。宣布開考後，考生須首先於適當位置貼上電腦條碼及填上各項所需資料。宣布停筆後，考生不會獲得額外時間貼上電腦條碼。

(二) 試場主任宣布開卷後，考生請檢查試題冊及確定試題冊內的試題。最後會有「**全卷完**」的字眼。

(三) 本試卷各題佔分相等。

(四) **本試卷全部試題均須回答**。為便於修正答案，考生宜用HB鉛筆把答案填畫在答題紙上。錯誤答案可用潔淨膠擦將筆痕徹底擦去。考生須清楚填畫答案，否則會因答案未能被辨認而失分。

(五) 每題只可填畫**一個**答案。如填劃超過一個答案，該題將**不獲評分**。

(六) 答案錯誤，不另扣分。

(七) 未經許多，請勿打開試題冊。

CC-CRE-APT

I. 演繹推理（8題）

請根據以下短文的內容，選出一個或一組推論。請假定短文的內容都是正確的。

1. **據報章在2017年報道，該年的1至10月吉林省工業實現利潤76.4億元，比前一年的同期增長近6倍。國有企業減虧15億元，減幅達42.2%；實現利潤67.4億元，增幅達8倍，這兩項指標均居全國前列。**

   據此，我們知道，根據當年1至10月的統計：

   A. 吉林省國有企業經已實現整體扭虧為盈

   B. 吉林省國有企業尚未實現整體扭虧為盈

   C. 吉林省工業增長速度在全國名列前茅

   D. 吉林省在建立現代企業制度方面取得顯著進展

2. **為了解決某地區長期嚴重的鼠患，一家公司生產了一種售價為2,500元的激光捕鼠器，該產品的捕鼠效果及使用性能堪稱一流，廠家為推出此產品又做了廣泛的廣告宣傳。但結果是產品仍沒有銷路。**

   請推測這家公司開發該新產品失敗的最主要原因可能是：

   A. 未能令廣大消費者了解該產品的優點

   B. 忽略消費者的價格承受力

   C. 人們不需要捕鼠

   D. 人們沒聽説過這種產品

3. **「知名度」和「美譽度」反映了社會公眾對一個組織的認識和贊許的程度，兩者都是公共關係學所強調追求的目標。一個組織形象如何，取決於它的知名度和美譽度。公共關係策劃者需要明白的是：只有不斷提高知名度，才能不斷提高組織的美譽度。知名度只有以美譽度為基礎才能產生積極的效應。同時，美譽度要以知名度為條件，才能充分顯示其社會價值。**
由此可知，知名度和美譽度的關係是：

A. 知名度高，美譽度必然高

B. 知名度低，美譽度必然低

C. 只有美譽度高，知名度才能高

D. 只有知名度高，美譽度才能高

4. **信貸消費在一些經濟發達國家十分盛行，很多消費者通過預支他們尚未到手的收入，來滿足對住屋、汽車、家用電器等耐用消費品的需求。在消費信貸發達的國家中，人們的普遍觀念是：不能負債説明你的信譽差。**
如果上述論述為真，那麼必須以下列哪項為前提：

A. 在發達國家，消費信貸已成為商業銀行擴大經營、加強競爭的重要手段

B. 消費信貸於國於民都有利，國家可利用利率下調刺激消費者購買商品

C. 社會已建立起完備、嚴密的信用網絡，銀行可對信貸者的經濟狀況進行查詢和監督

D. 保險公司可向借貸人提供保險，以保障銀行資產的安全。

5. **由於近期的乾旱和高溫，導致海灣鹽度增加，引起了許多魚死亡。蝦雖然可以適應高鹽度，但鹽度高也給養蝦場帶來了不幸。**
以下哪項如果為真，能夠提供解釋以上現象的原因？

A. 持續的乾旱會使海灣的水位下降，這經已引起有關機構的注意

B. 幼蝦吃的有機物在鹽度高的環境下，幾乎難以存活

C. 水溫升高會使蝦更快速的繁殖

D. 魚多的海灣往往蝦也多，蝦少的海灣魚也不多

6. **某Z國生產汽車發動機的成本比Y國低10%，即使加上關稅和運輸費，從Z國進口汽車發動機仍比在Y國生產便宜。**
由此我們可以知道：

A. Z國的勞動力成本比Y國低10%

B. 從Z國進口汽車發動機的關稅低於在Y國生產成本的10%

C. 由Z國運一個汽車發動機的費用高於在Y國造一個汽車發動機的10%

D. 由Z國生產一個汽車發動機的費用是Y國的10%

7. **世界食品需求能否保持平衡，取決於人口和經濟增長的速度。人口增長會導致食物攝取量的增加；另一方面，經濟增長會促使畜產品消費增加，改變人們的食物結構，從而對全球的穀物需求產生影響。**

據此可知：

A. 人口的增長將影響全球的穀物需求

B. 改變食物結構將降低全球的穀物需求

C. 經濟的增長可降低全球穀物的需求

D. 人口的增長會導致世界畜產品消費的增加

8. **一方面由於天氣惡劣，另一方面由於主要的胡椒種植者轉種高價位的可可，所以，三年以來，全世界的胡椒產量已經遠遠低於銷售量了。因此，目前胡椒的供應相當短缺。其結果是：胡椒價格上揚，已經和可可不相上下了。**

據此可知：

A. 世界市場胡椒的儲存量正在減少

B. 世界胡椒消費已持續三年居高不下

C. 胡椒種植者正在擴大胡椒的種植面積

D. 目前可可的價格要低於三年前的價格

II. Verbal Reasoning (English) (6 questions)

Directions :
In this test, each passage is followed by three statements (the questions). You have to assume what is stated in the passage is true and decide whether the statements are either:
True (Box A): the statement is already made or implied in the passage, or follows logically from the passage.
False (Box B): the statement contradicts what is said, implied by, or follows logically from the passage.
Can't tell (Box C): there is insufficient information in the passage to establish whether the statement is true or false.

Passage 1 (Question 9 to 11):

Being socially responsible is acting ethically and showing integrity. It directly affects our quality of life through such issues as human rights, working conditions, the environment, and corruption. It has traditionally been the sole responsibility of governments to police unethical behaviour. However, the public have realised the influence of corporations and, over the last ten years, the level of voluntary corporate social responsibility initiatives that dictate the actions of corporations has increased.

9. The ethical actions of corporations have changed over the last ten years.

10. Corporations can influence the public's quality of life.

11. Traditionally, the government have relied upon only the large corporations to help drive corporate social responsibility, whilst they concentrated on the smaller corporations.

Passage 2 (Question 12 to 14):

Stem cells are cells that can self-renew and differentiate into specialised cell types. Because of their potential to replace diseased or defective human tissue, stem cells are seen by scientists as key to developing new therapies for a wide range of conditions, including degenerative illnesses and genetic diseases. Treatments based on adult stem cells – from sources such as umbilical cord blood – have been successfully developed, but what makes stem cell research controversial is the use of embryonic stem cells. Not only do embryonic stem cells reproduce more quickly than adult stem cells, they also have wider differentiation potential. The main opponents to stem cell research are pro-life supporters, who believe that human life should not be destroyed for science. Advocates of stem cell research counter this crucial point by saying that an embryo cannot be viewed as a human life, and that medical advances justify the destruction of embryos. Furthermore, stem cell research utilises the thousands of surplus embryos created for in vitro fertilisation treatment. The issue is particularly divisive in the United States, where federal funding is not available for the creation of new embryonic stem cell lines, although recent legislation has opened up government funding to further research on embryonic stem cells created through private funding. Whereas many governments prohibit the production of embryonic stem cells, it is allowed in countries including the UK, Sweden and the Netherlands.

12. Stem cells are at the forefront of medical research because of their ability to grow indefinitely.

13. The United States government does not supply funding for projects using embryonic stem cell lines.

14. One advantage of embryonic stem cells over adult stem cells is their greater ability to be converted into specialised cell types.

III. Data Sufficiency Test (8 questions)

Directions : In this test, you are required to choose a combination of clues to solve a problem.

15. How is ‘M’ related to ‘N’ ?
    (1) ‘P’ is the daughter of ‘M’ and mother of ‘S’
    (2) ‘T’ is the son of ‘P’ and husband of ‘N’

    A. If the data in statement (1) alone are sufficient

    B. If the data in statement (2) alone are sufficient

    C. If the data either in statement (1) alone or in statement (2) alone are sufficient

D. If the data given in both the statements (1) and (2) together are not sufficient

E. If the data in both the statements (1) and (2) together are necessary

16. What is the monthly salary of Praveen?

(1) Praveen gets 15% more than Sumit while Sumit gets 10% less than Lokesh

(2) Lokesh's monthly salary is Rs.2500

A. if the data in statement (1) alone are sufficient to answer the question

B. if the data in statement (2) alone are sufficient answer the question

C. if the data either in (1) or (2) alone are sufficient to answer the question

D. if the data even in both the statements together are not sufficient to answer the question

E. If the data in both the statements together are needed

17. How is ‘face’ written in that code language?

(1) In a certain code language, ‘no one with face’ is coded as ‘fo to om sop’ and ‘no one has face’ is coded as ‘om sit fo sop’

(2) In a certain code language, ‘face of no light’ is coded as ‘om mot fo kiz’ and ‘no one is smart’ is coded as ‘sop fo sip lik’.

A. If data in the statement (1) alone is sufficient to answer the question

B. If data in the statement (2) alone is sufficient to answer the question

C. If data either in the statement (1) alone or statement (2) alone are sufficient to answer the question

D. If data given in both (1) & (2) together are not sufficient to answer the question

E. If data in both statements (1) & (2) together are necessary to answer the question

18. On which day in April is Gautham’ s birthday?

(1) Gautam was born exactly 28 years after his mother was born

(2) His mother will be 55 years 4 months and 5 days on August 18 this year

A. if the data in statement (1) alone are sufficient to answer the question

B. if the data in statement (2) alone are sufficient answer the question

C. if the data either in (1) or (2) alone are sufficient to answer the question

D. if the data even in both the statements together are not sufficient to answer the question

E. If the data in both the statements together are needed

19. Who has secured less marks among P, Q, R , S & T?

(1) S has secured less marks than only R and T

(2) Q secured more marks than P

A. If data in the statement (1) alone is sufficient to answer the question

B. If data in the statement (2) alone is sufficient to answer the question

C. If data either in the statement (1) alone or statement (2) alone are sufficient to answer the question

D. If data given in both (1) & (2) together are not sufficient to answer the question

E. If data in both statements (1) & (2) together are necessary to answer the question

20. How many boys students are there in the class?

(1) 65% girls students are there in the class

(2) The no. of boys students is half that of girls

A. If data in the statement (1) alone is sufficient to answer the question.

B. If data in the statement (2) alone is sufficient to answer the question.

C. If data either in the statement (1) alone or statement (2) alone are sufficient to answer the question.

D. If data given in both (1) & (2) together are not sufficient to answer the question.

E. If data in both statements (1) & (2) together are necessary to answer the question.

21. In a certain code, 'nop al ed' means 'They like flowers'. Which code word means 'flowers'?

(1) 'id nim nop' means 'They are innocent'

(2) 'gob ots al' means 'We like roses'

A. If the data in statement (1) alone are sufficient

B. If the data in statement (2) alone are sufficient

C. If the data either in statement (1) alone or in statement (2) alone are sufficient

D. If the data given in both the statements (1) and (2) together are not sufficient

E. If the data in both the statements (1) and (2) together are necessary

22. What is Meena' s rank from top in a class room of twenty students?

(1) Rama is fifth from the top and two ranks above Meena

(2) Ashok is tenth from the bottom and three ranks below Meena

A. If the data in statement (1) alone are sufficient

B. If the data in statement (2) alone are sufficient

C. If the data either in statement (1) alone or in statement (2) alone are sufficient

D. If the data given in both the statements (1) and (2) together are not sufficient

E. If the data in both the statements (1) and (2) together are necessary

IV. Numerical Reasoning (5 questions)

Directions :

Each question is a sequence of numbers with one or two numbers missing. You have to figure out the logical order of the sequence to find out the missing number(s).

23. 1/2，1，7/6，5/4，13/10，( )

A. 4/3

B. 3/4

C. 11/14

D. 17/18

24. [(9，6) 42 (7，7)] [(7，3) 40 (6，4)] [(8，2) ( ) (3，2)]

A. 30

B. 32

C. 34

D. 36

25. 1，1，8/7，16/11，2，( )

A. 36/23

B. 9/7

C. 32/11

D. 35/22

26. 35，7，5，(　)，25/7

A. 1

B. 7/5

C. 3

D. 5/7

27. 1.01，1.02，2.03，3.05，5.08，(　)

A. 8.13

B. 8.013

C. 7.12

D. 7.012

V. Interpretation of Tables and Graphs (8 questions)

Directions :
This is a test on reading and interpretation of data presented in tables and graphs.

The following pie-charts show the distribution of students of graduate and post-graduate levels in seven different institutes in a town.

Chart 1

Distribution of students at graduate and post-graduate levels in seven institutes:

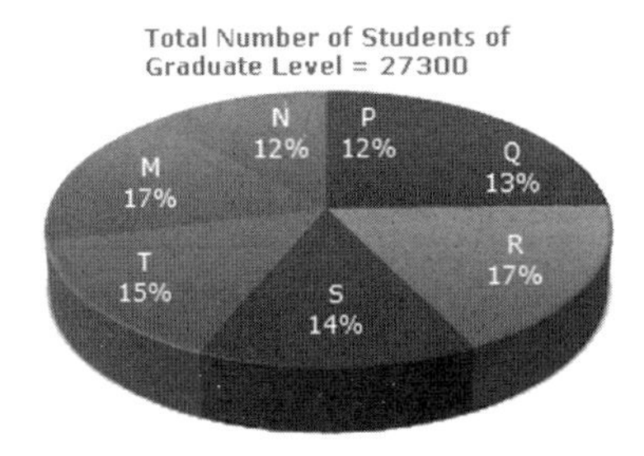

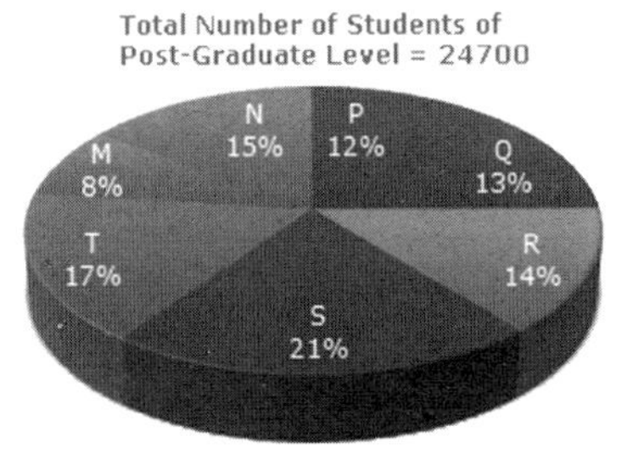

28. What is the total number of graduate and post-graduate level students is institute R ?

    A. 8320

    B. 7916

    C. 9116

    D. 8099

29. What is the ratio between the number of students studying at post-graduate and graduate levels respectively from institute S?

A. 14: 19

B. 19:21

C. 17:21

D. 19:14

30. How many students of institutes of M and S are studying at graduate level?

A. 7516

B. 8463

C. 9127

D. 9404

31. What is the ratio between the number of students studying at post-graduate level from institutes S and the number of students studying at graduate level from institute Q?

A. 13:19

B. 21:13

C. 13:8

D. 19:13

Chart 2

The following line graph gives the percentage of the number of candidates who qualified an examination out of the total number of candidates who appeared for the examination over a period of seven years from 1994 to 2000.

Percentage of Candidates Qualified to Appeared in an Examination Over the Years

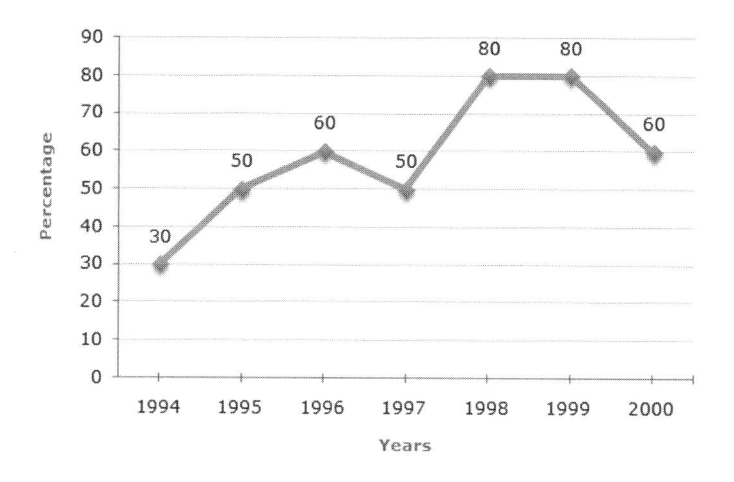

32. The difference between the percentage of candidates qualified to appeared was maximum in which of the following pairs of years?

 A. 1994 and 1995

 B. 1997 and 1998

 C. 1998 and 1999

 D. 1999 and 2000

33. In which pair of years was the number of candidates qualified, the same?

A. 1995 and 1997

B. 1995 and 2000

C. 1998 and 1999

D. Data inadequate

34. If the number of candidates qualified in 1998 was 21200, what was the number of candidates appeared in 1998?

A. 32000

B. 28500

C. 26500

D. 25000

35. If the total number of candidates appeared in 1996 and 1997 together was 47400, then the total number of candidates qualified in these two years together was?

A. 34700

B. 32100

C. 31500

D. Data inadequate

全卷完

CRE-APT

文化會社出版社 **CULTURE CROSS LIMITED**

答題紙 ANSWER SHEET

請在此貼上電腦條碼
Please stick the barcode label here

(1) 考生編號 Candidate No.

(2) 考生姓名 Name of Candidate

(3) 考生簽署 Signature of Candidate

宜用H.B.鉛筆作答
You are advised to use H.B. Pencils

考生須依照下圖所示填畫答案：

23 A B C D E

錯填答案可使用潔淨膠擦將筆痕徹底擦去。

切勿摺皺此答題紙

Mark your answer as follows:

23 A B C D E

Wrong marks should be completely erased with a clean rubber.

DO NOT FOLD THIS SHEET

1	A	B	C	D	E	21	A	B	C	D	E
2	A	B	C	D	E	22	A	B	C	D	E
3	A	B	C	D	E	23	A	B	C	D	E
4	A	B	C	D	E	24	A	B	C	D	E
5	A	B	C	D	E	25	A	B	C	D	E
6	A	B	C	D	E	26	A	B	C	D	E
7	A	B	C	D	E	27	A	B	C	D	E
8	A	B	C	D	E	28	A	B	C	D	E
9	A	B	C	D	E	29	A	B	C	D	E
10	A	B	C	D	E	30	A	B	C	D	E
11	A	B	C	D	E	31	A	B	C	D	E
12	A	B	C	D	E	32	A	B	C	D	E
13	A	B	C	D	E	33	A	B	C	D	E
14	A	B	C	D	E	34	A	B	C	D	E
15	A	B	C	D	E	35	A	B	C	D	E
16	A	B	C	D	E	36	A	B	C	D	E
17	A	B	C	D	E	37	A	B	C	D	E
18	A	B	C	D	E	38	A	B	C	D	E
19	A	B	C	D	E	39	A	B	C	D	E
20	A	B	C	D	E	40	A	B	C	D	E

文化會社出版社
投考公務員 模擬試題王

能力傾向試測試
模擬試卷（四）

時間：四十五分鐘

考生須知：

(一) 細讀答題紙上的指示。宣布開考後，考生須首先於適當位置貼上電腦條碼及填上各項所需資料。宣布停筆後，考生不會獲得額外時間貼上電腦條碼。

(二) 試場主任宣布開卷後，考生請檢查試題冊及確定試題冊內的試題。最後會有「**全卷完**」的字眼。

(三) 本試卷各題佔分相等。

(四) **本試卷全部試題均須回答**。為便於修正答案，考生宜用HB鉛筆把答案填畫在答題紙上。錯誤答案可用潔淨膠擦將筆痕徹底擦去。考生須清楚填畫答案，否則會因答案未能被辨認而失分。

(五) 每題只可填畫**一個**答案。如填劃超過一個答案，該題將**不獲評分**。

(六) 答案錯誤，不另扣分。

(七) 未經許多，請勿打開試題冊。

CC-CRE-APT

I. 演繹推理（8題）

請根據以下短文的內容，選出一個或一組推論。請假定短文的內容都是正確的。

1. **在公元之初，牧師們將基督教在整個羅馬帝國中傳播開來。基督教的普及首先涉及的是各大城市的平民階層。基督徒們反對皇權意識，因此君主們都把他們視為威脅，並組織起對他們的暴力摧殘。然而這並沒有中斷新信仰的擴展，它一點點贏得了貴族階級和最有影響力的人士，而且很快觸及到了皇帝周圍的人和君主本人。公元312年的米蘭敕令為迫害畫上了句號。公元380年，狄奧多西赦令使基督教成為國教。因此，公元4世紀標誌著基督教概念中的轉捩點。**

   根據這段話，可得到的正確推論是：

   A. 基督教從開始就反對皇權統治，且一直與羅馬皇帝及其政權鬥爭

   B. 基督教成為國教，意味著它由非法而合法、由被迫害而被推崇

   C. 公元312年基督教贏得了貴族階級和最有影響力的人士的支持

   D. 公元380年以後，基督徒不再受到迫害，基督教被承認

2. **作為唯一一支留在世界盃的南美球隊，下一場比賽巴西將迎戰淘汰了丹麥的英格蘭球隊。巴西隊教練斯史高拉利不願談論如何與英格蘭較量，而他的隊員也保持著清醒的頭腦。在擊敗頑強的比利時隊後，史高拉利如釋重負：「我現在腦子裡想的第一件事就是好好放鬆一下。」**

依據上文，我們無法推知的是：

A. 巴西隊在本屆世界盃中再也不會與南美球隊比賽

B. 由於沒有做好充分的準備，史高拉利不願意談論與英格蘭的較量

C. 與比利時的比賽很艱苦，所以賽後史高拉利如釋重負

D. 英格蘭在與巴西比賽之前必須要戰勝丹麥

3. **四個小偷(每人各偷了一種東西)接受盤問。甲說：「每人只偷了一隻錶」；乙說：「我只偷了一顆鑽石」；丙說：「我沒偷錶」；丁說：「有些人沒偷錶」。經過警察的進一步調查，發現這次審問中只有一人說真話。**
下列判斷，沒有失誤的是：

A. 所有人都偷了錶

B. 所有人都沒偷錶

C. 有些人沒偷錶

D. 乙偷了一顆鑽石

4. 為了胎兒的健康，孕婦一定要保持身體健康。為了保持身體健康，她必須攝取足夠的鈣質，同時，為了攝取到足夠的鈣質，她必須喝牛奶。
據此可知：

A. 如果孕婦不喝牛奶，胎兒就會發育不好

B. 攝取了足夠的鈣質，孕婦就會身體健康

C. 孕婦喝牛奶，她就會身體健康

D. 孕婦喝牛奶，胎兒就會發育良好

5. 張老師的班裡有60個學生，男、女生各佔一半。有40個學生喜歡數學；有50個學生喜歡語文。
這表明可能會：

A. 有20個男生喜歡數學而不喜歡語文

B. 有20個喜歡語文的男生不喜歡數學

C. 有30個喜歡語文的女生不喜歡數學

D. 有30個喜歡數學的男生只有10個喜歡語文

6. 實驗發現少量口服某種安定藥，可使人們在測謊的測驗中撒謊而不被發現。測謊所產生的心理壓力能夠被這類安定藥物有效地抑制，同時沒有顯著的副作用。因此，這類藥物可同時有效地減少日常生活的心理壓力而無顯著的副作用。以下哪項最可能是題幹的論證所假設？

A. 任何類型的安定藥物都有抑制心理壓力的效果

B. 如果禁止測試者服用任何藥物，測謊器就有完全準確的測試結果

C. 測謊所產生的心理壓力與日常生活人們面臨的心理壓力類似

D. 越來越多的人在日常生活中面臨日益加重的心理壓力

7. **某國連續四年的統計表明，在夏令時改變的時間裡比其他時間的車禍高4%。這些統計結果説明時間的改變嚴重影響了某國司機的注意力。**
得出這一結論的前提條件是：

A. 該國的司機和其他國家的司機有相同的駕駛習慣

B. 被觀察到的事故率的增加幾乎都是歸因於小事故數量的增加

C. 關於交通事故發生率的研究，至少需要5年的觀察

D. 沒有其他的諸如學校假期和節假日導致車禍增加的因素

8. **當一個人處於壓力下的時候，他更可能得病。**
下列説法最能支持上述結論的是：

A. 研究顯示，處於醫院或診所中是一個有壓力的環境

B. 許多企業反映職員在感到工作壓力增大時，缺勤明顯減少

C. 在放假期間，大學醫院的就診人數顯著增加

D. 在考試期間，大學醫院的就診人數顯著增加

II. Verbal Reasoning (English) (6 questions)

Directions :
In this test, each passage is followed by three statements (the questions). You have to assume what is stated in the passage is true and decide whether the statements are either:
True (Box A): the statement is already made or implied in the passage, or follows logically from the passage.
False (Box B): the statement contradicts what is said, implied by, or follows logically from the passage.
Can't tell (Box C): there is insufficient information in the passage to establish whether the statement is true or false.

Passage 1 (Question 9 to 11):

The project was ambitious in its size, complexity, triparty nature, and in its pioneering of the Private Finance Initiative. This difficulty was unavoidable and contributed to the project's failure. However, a more thorough estimate of the unknown difficulties and timescales would have enabled the Department to better prepare for the project, and increase its chance of success.
In December 1997 XSoft indicated they needed time to complete the project, which should have been inevitable. If the Department knew from the start how long the project would take, it is questionable whether they would have considered inception, especially considering the implications of delay on the overall profitability for the venture.

9. If more care had been put into estimating the difficulties, it is less likely the project would have failed.

10. XSoft witheld information from the Department regarding how long the project would take.

11. The Department' s profits were dependent upon how long the project took.

Passage 2 (Question 12 to 14):

An economic bubble is a situation in which the price of an asset is greatly inflated to its intrinsic value. Famous examples of bubbles include the Dutch tulip mania of the 1600s, the Dot-com bubble of the late 1990s and early 2000s and the recent housing bubble in various countries around the world. Just like on many other economic issues, economists do not have a consensus over the causes of, effects of, or solutions to bubbles. Oft cited causes of recent bubbles include the easy/cheap availability of credit (owing in part to loose monetary policy), greater fool theory, moral hazard and herding behaviour. Most economists do agree that bubbles can lead to the misallocation of scarce resources (which is the central question of economics). Some economists believe government ought to step in to 'pop' a bubble through contractionary fiscal or monetary policy, while others believe the bubble should be allowed to run its course and deflate naturally.

12. There is consensus among economists over the solutions to bubbles.

13. Some causes of bubbles may be loose monetary policy and moral hazard.

14. The central question of economics is the allocation of scarce resources.

III. Data Sufficiency Test (8 questions)

Directions : In this test, you are required to choose a combination of clues to solve a problem.

15. How is 'go' written in a code language?

(1) 'you may go' is written as 'pit ja ho' in that code language.

(2) 'he may come' is written as 'ja da na' in that code language.

A. If the data in statement (1) alone are sufficient

B. If the data in statement (2) alone are sufficient

C. If the data either in statement (1) alone or in statement (2) alone are sufficient

D. If the data given in both the statements (1) and (2) together are not sufficient

E. If the data in both the statements (1) and (2) together are necessary

16. At what time did Sonali leave her home for office?

(1) Sonali received a phone call at 9:15 a.m. at her home

(2) Sonali' s car reached office at 10:15 a.m., 45 minutes after she left her residence

A. if the data in statement (1) alone are sufficient to answer the question

B. if the data in statement (2) alone are sufficient answer the question

C. if the data either in (1) or (2) alone are sufficient to answer the question

D. if the data even in both the statements together are not sufficient to answer the question

E. If the data in both the statements together are needed

17. What is the shortest distance between Devipur and Durgapur?

(1) Durgapur is 20 km away from Rampur

(2) Devipur is 15 km away from Rampur

A. If the data in statement (1) alone are sufficient

B. If the data in statement (2) alone are sufficient

C. If the data either in statement (1) alone or in statement (2) alone are sufficient

D. If the data given in both the statements (1) and (2) together are not sufficient

E. If the data in both the statements (1) and (2) together are necessary

18. Kiran is older than Manoj and Dilip is older than Neelam. Who among them is the youngest?

(1) Kiran is older than Neelam

(2) Manoj is younger than Dilip

A. If the data in statement (1) alone are sufficient

B. If the data in statement (2) alone are sufficient

C. If the data either in statement (1) alone or in statement (2) alone are sufficient

D. If the data given in both the statements (1) and (2) together are not sufficient

E. If the data in both the statements (1) and (2) together are necessary

19. A, B, C and D made their project presentation, one on each day, on four consecutive days but not necessarily in that order. On which day did ‘C’ make his presentation?

(1) The first presentation was made on 23rd, Tuesday and was followed by D’s presentation.

(2) ‘A’ did not make his presentation on 25th and one of them made his presentation, between A’s and B’s.

A. If the data in statement (1) alone are sufficient

B. If the data in statement (2) alone are sufficient

C. If the data either in statement (1) alone or in statement (2) alone are sufficient

D. If the data given in both the statements (1) and (2) together are not sufficient

E. If the data in both thestatements (1) and (2) together are necessary

20. What is the amount of rice exported from India?

(1) India' s export to America is 80,000 tonnes and this is 10% of the total rice exports.

(2) India' s total export tonnage of rice is 12.5% of the total of 1.9 million tonnes.

A. if the data in statement (1) alone are sufficient to answer the question

B. if the data in statement (2) alone are sufficient answer the question

C. if the data either in (1) or (2) alone are sufficient to answer the question

D. if the data even in both the statements together are not sufficient to answer the question

E. If the data in both the statements together are needed

21. Gourav ranks eighteenth in a class. What is his rank from the last?

(1) There are 47 students in the class

(2) Jatin who ranks 10th in the same class, ranks 38th from the last

A. if the data in statement (1) alone are sufficient to answer the question

B. if the data in statement (2) alone are sufficient answer the question

C. if the data either in (1) or (2) alone are sufficient to answer the question

D. if the data even in both the statements together are not sufficient to answer the question

E. If the data in both the statements together are needed

22. On which day of the week did Sohan visit Chennai?

(1) Sohan returned to Hyderabad from Chennai on Thursday

(2) Sohan left Hyderabad on Monday for Chennai

A. If the data in statement (1) alone are sufficient

B. If the data in statement (2) alone are sufficient

C. If the data either in statement (1) alone or in statement (2) alone are sufficient

D. If the data given in both the statements (1) and (2) together are not sufficient

E. If the data in both the statements (1) and (2) together are necessary

IV. Numerical Reasoning (5 questions)

Directions :

Each question is a sequence of numbers with one or two numbers missing. You have to figure out the logical order of the sequence to find out the missing number(s).

23. 28，16，12，4，8，( )

A. -8

B. 6

C. -4

D. 2

24. 5，16，29，45，66，94，( )

A. 114

B. 121

C. 133

D. 142

25. 29，21，15，( )，9

A. 17

B. 11

C. 25

D. 7

26. 0，3，9，21，(　)，93

A. 40

B. 45

C. 36

D. 38

27. 3，8，22，62，178，(　)

A. 518

B. 516

C. 548

D. 546

V. Interpretation of Tables and Graphs (8 questions)

Directions :
This is a test on reading and interpretation of data presented in tables and graphs.

Chart 1

A school has four sections A, B, C, D of Class IX students.
The results of half yearly and annual examinations are shown in the table given below.

Result	No. of Students			
	Section A	Section B	Section C	Section D
Students failed in both Exams	28	23	17	27
Students failed in half-yearly but passed in Annual Exams	14	12	8	13
Students passed in half-yearly but failed in Annual Exams	6	17	9	15
Students passed in both Exams	64	55	46	76

28. If the number of students passing an examination be considered a criteria for comparison of difficulty level of two examinations, which of the following statements is true in this context?

    A. Half yearly examinations were more difficult.

    B. Annual examinations were more difficult.

    C. Both the examinations had almost the same difficulty level.

    D. The two examinations cannot be compared for difficulty level.

29. How many students are there in Class IX in the school?

A. 336

B. 189

C. 335

D. 430

30. Which section has the maximum pass percentage in at least one of the two examinations?

A. A Section

B. B Section

C. C Section

D. D Section

31. Which section has the maximum success rate in annual examination?

A. A Section

B. B Section

C. C Section

D. D Section

Chart 2

The following bar chart shows the trends of foreign direct investments(FDI) into India from all over the world.

Trends of FDI in India

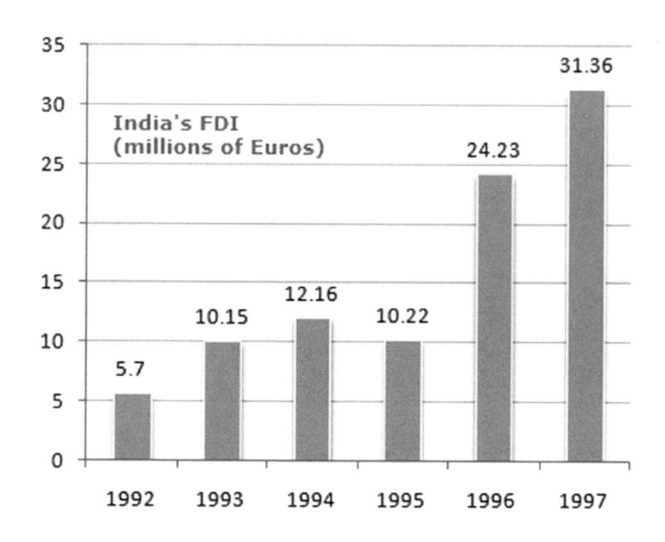

32. What was the ratio of investment in 1997 over the investment in 1992?

A. 5.50

B. 5.36

C. 5.64

D. 5.75

33. What was absolute difference in the FDI to India in between 1996 and 1997?

A. 7.29

B. 7.13

C. 8.13

D. None of these

34. If India FDI from OPEC countries was proportionately the same in 1992 and 1997 as the total FDI from all over the world and if the FDI in 1992 from the OPEC countries was Euro 2 million. What was the amount of FDI from the OPEC countries in 1997?

A. 11

B. 10.72

C. 11.28

D. 11.5

35. Which year exhibited the highest growth in FDI in India over the period shown?

A. 1993

B. 1994

C. 1995

D. 1996

全卷完

CRE-APT

文化會社出版社 **CULTURE CROSS LIMITED**

答題紙 ANSWER SHEET

請在此貼上電腦條碼
Please stick the barcode label here

(1) 考生編號 Candidate No.

(2) 考生姓名 Name of Candidate

(3) 考生簽署 Signature of Candidate

宜用H.B.鉛筆作答
You are advised to use H.B. Pencils

考生須依照下圖所示填畫答案：

23 A B C D E

錯填答案可使用潔淨膠擦將筆痕徹底擦去。

切勿摺皺此答題紙

Mark your answer as follows:

23 A B C D E

Wrong marks should be completely erased with a clean rubber.

DO NOT FOLD THIS SHEET

1	A	B	C	D	E	21	A	B	C	D	E
2	A	B	C	D	E	22	A	B	C	D	E
3	A	B	C	D	E	23	A	B	C	D	E
4	A	B	C	D	E	24	A	B	C	D	E
5	A	B	C	D	E	25	A	B	C	D	E
6	A	B	C	D	E	26	A	B	C	D	E
7	A	B	C	D	E	27	A	B	C	D	E
8	A	B	C	D	E	28	A	B	C	D	E
9	A	B	C	D	E	29	A	B	C	D	E
10	A	B	C	D	E	30	A	B	C	D	E
11	A	B	C	D	E	31	A	B	C	D	E
12	A	B	C	D	E	32	A	B	C	D	E
13	A	B	C	D	E	33	A	B	C	D	E
14	A	B	C	D	E	34	A	B	C	D	E
15	A	B	C	D	E	35	A	B	C	D	E
16	A	B	C	D	E	36	A	B	C	D	E
17	A	B	C	D	E	37	A	B	C	D	E
18	A	B	C	D	E	38	A	B	C	D	E
19	A	B	C	D	E	39	A	B	C	D	E
20	A	B	C	D	E	40	A	B	C	D	E

文化會社出版社
投考公務員 模擬試題王

能力傾向試測試
模擬試卷（五）

時間：四十五分鐘

考生須知：

(一) 細讀答題紙上的指示。宣布開考後，考生須首先於適當位置貼上電腦條碼及填上各項所需資料。宣布停筆後，考生不會獲得額外時間貼上電腦條碼。

(二) 試場主任宣布開卷後，考生請檢查試題冊及確定試題冊內的試題。最後會有「**全卷完**」的字眼。

(三) 本試卷各題佔分相等。

(四) **本試卷全部試題均須回答**。為便於修正答案，考生宜用HB鉛筆把答案填畫在答題紙上。錯誤答案可用潔淨膠擦將筆痕徹底擦去。考生須清楚填畫答案，否則會因答案未能被辨認而失分。

(五) 每題只可填畫**一個**答案。如填劃超過一個答案，該題將**不獲評分**。

(六) 答案錯誤，不另扣分。

(七) 未經許多，請勿打開試題冊。

CC-CRE-APT

I. 演繹推理（8題）

請根據以下短文的內容，選出一個或一組推論。請假定短文的內容都是正確的。

1. **某國的人口調查數據表明，三十歲以上的未婚男性超過這個年齡段的未婚女性，比例大約為10:1。這些男性中，大多數確實希望結婚。然而，明顯地，除非他們中有許多人娶外國女性，否則大部分人將保持未婚。**

   上面論述基於下列哪個假設？

   A. 移民到外國的女性比男性多

   B. 與該國男性相同年齡的三十幾歲的該國女性寧願保持未婚

   C. 大多數未婚的該國男性不可能娶比他們年齡大的女性

   D. 大多數未婚的該國男性不願意娶非本國的女性

2. **根據2018年度的統計顯示，對中國人的健康威脅的三種慢性病，按其在總人口中的發病率排列，依次是乙型肝炎、關節炎和高血壓。其中，關節炎和高血壓的發病率隨著年齡的增長而增加，而乙型肝炎在各個年齡段的發病率沒有明顯的不同。中國人口的平均年齡，在2018年至2030年之間，將呈明顯上升態勢而逐步進入老齡化社會。**

   根據題幹提供的信息，推出以下哪項結論最為恰當？

   A · 到2030年，發病率的將是關節炎

   B · 到2030年，發病率的將仍是乙型肝炎

   C · 在2018年至2030年之間，乙型肝炎患者的平均年齡將增大

   D · 到2030年，乙型肝炎的老年患者將多於非老年患者

3. **下列推理哪一項是正確的：**

A. 只有開啟電源開關，日光燈才亮，這盞日光燈不亮，所以沒有開啟電源開關

B. 只有意思表示真實的行為，才屬於民事法律行為，張、王二人的借貸行為是意思表示真實的行為，所以，他們之間的借貸行為屬於民事法律行為

C. 如果黃某是本案犯案人，那麼他就具有犯案工具；如果黃某是本案犯案人，那麼他就有犯案時間。黃某既不具有犯案工具，也沒有犯案時間，所以，黃某不是本案犯案人

D. 犯案人是熟悉現場情況的人；這個公司的人是熟悉現場情況的人，所以，這個公司的人是犯案人

4. **有時為了醫治一些危重病人，醫院允許使用海洛英作為止痛藥。其實，這樣做是應當禁止的。因為，毒品販子會這種渠道獲取海洛英，這會對社會造成嚴重危害。**
以下哪個如果為真，最能削弱以上的論證？

A. 有些止痛藥可以起到和海洛英一樣的止痛效果

B. 用於止痛的海洛英在數量上與用於做非法交易的比起來，是微不足道的

C. 海洛英如果用量過大就會致死

D. 在治療過程中，海洛英的使用不會使病人上癮

5. **由「高薪未必養廉」這句話我們可以推出：**

A. 低薪同樣養廉

B. 養廉必定高薪

C. 高薪難以養廉

D. 低薪肯定養廉

6. **林先生和張先生參選議員選舉。在選前10天進行的民意測驗顯示，36%的受訪者打算選林先生，42%打算選張先生。而在最後的正式選舉中，林先生的得票率是52%，張先生的得票率有46%。這説明選舉前的民意測驗的操作上出現了失誤。**
以下哪項如果是真的，最能削弱上述論證的結論?

A. 選舉前20天進行的民意測驗顯示，林先生的得票率是32%，張先生的得票率是40%

B. 在進行民意測驗的時候，許多選舉者還沒拿定主意選誰

C. 在選舉的前七天，林先生為廠裡要回了30萬元借款

D. 林先生在競選中的演説能力要比張先生強

7. **不久前，還有許多人從事鉛字排印、電報發送和機械打字工作，也有一些人是這些職業領域中的技術高手。今天，這些職業已經從社會上消失了。由於基因技術的發展，可能會幫助人類解決「近視」的問題，若干年後，今天非常興旺的眼鏡行業也可能會趨於消失。**
據此，我們知道：

A. 一些新的職業會誕生

B. 一些人的職業變換與技術發展有關

C. 今後，許多人在一生中會至少從事兩種以上的職業

D. 終生教育是未來教育發展的大趨勢

8. **有人認為，一個國家如果能有效率地運作經濟，就一定能創造財富而變得富有；而這樣的一個國家要想保持政治穩定，它所創造的財富必須得到公正的分配；而財富的公正分配將結束經濟風險；但是，風險的存在正是經濟有效率運作的不可或缺的先決條件。**

從上述觀點可以得出以下哪項結論?

A. 一個國家政治上的穩定和經濟上的富有不可能並存

B. 一個國家政治上的穩定和經濟上的有效率運作不可能並存

C. 一個富有國家的經濟運作一定是有效率的

D. 一個政治上不穩定的國家，一定同時充滿了經濟風險

II. Verbal Reasoning (English) (6 questions)

Directions :

In this test, each passage is followed by three statements (the questions). You have to assume what is stated in the passage is true and decide whether the statements are either:

True (Box A): the statement is already made or implied in the passage, or follows logically from the passage.

False (Box B): the statement contradicts what is said, implied by, or follows logically from the passage.

Can't tell (Box C): there is insufficient information in the passage to establish whether the statement is true or false.

Passage 1 (Question 9 to 11):

A well-nourished child can be more likely to be a studious one. But food has been seen as a cost to be cut in times of austerity, rather than an ingredient of good schooling. That may now be changing: as the government worries about obesity – which is fast rising among children- and urges everyone to eat less salt, fat and sugar, and more fruit and vegetables, the deficiency and unhealthiness of most school meals is striking. But cash constraints make change difficult.

9. Children who eat healthily will perform better in exams.

10. The number of obese children used to be less than it is now.

11. The government is apathetic about obesity.

Passage 2 (Question 12 to 14):

In biology, the term mutualism refers to a mutually beneficial relationship between two species. The later economic theory of mutualism is based on the labour theory of value, which states that the true cost of an item is the amount of labour that was required to produce it. Hence, goods should not be sold for more than the cost of acquiring them. Mutualism is closely associated with anarchism, because its principles were set forth in the mid- nineteenth century by the French politician and philosopher Pierre-Joseph Proudhon – the first person to define himself as an "anarchist". The main tenets of mutualism are free association and free credit. In a mutualist workplace, workers with different skills form an association to create a product or service. Though dependent on each other, the workers are not subordinated as in a capitalist enterprise. Mutual banks, also called credit unions, operate on the belief that free credit enables profit to be generated for the benefit of the union's members rather than bankers. Modern-day mutualism is sometimes described as free-market socialism. Proponents of mutualism support a free market economy, but object to capitalism because of the inequalities created by government intervention. Many mutual businesses and banking establishments exist today, espousing Proudhon's Co-operative model.

12. Proudhon’s economic theory of mutualism was influenced by biological mutualism.

13. Mutual banking establishments do not operate on a for-profit basis.

14. The labour theory of value is defined as: only the person who made an item should profit from its sale.

III. Data Sufficiency Test (8 questions)

Directions : In this test, you are required to choose a combination of clues to solve a problem.

15. How many candidates were interviewed everyday by the panel ‘A’ out of the three panels A, B and C?

    (1) The three panels on an average can intervies 15 candidates every day

    (2) Out of a total of 45 candidates interviewed everyday by the three panels, the no.of candidates interviewed by panel ‘A’ is more by 2 than the candidates interviewed by panel ‘C’ and is less by 1 than the candidates interviewed by panel ‘B’

A. If the data in statement (1) alone are sufficient

B. If the data in statement (2) alone are sufficient

C. If the data either in statement (1) alone or in statement (2) alone are sufficient

D. If the data given in both the statements (1) and (2) together are not sufficient

E. If the data in both the statements (1) and (2) together are necessary

16. Brinda' s merit rank is 17th in her class. What is the rank from the last?

(1) There are 70 students in her class

(2) Nisha who ranks 20th in Brinda' s class is 51st from the last

A. If the data in statement (1) alone are sufficient

B. If the data in statement (2) alone are sufficient

C. If the data either in statement (1) alone or in statement (2) alone are sufficient

D. If the data given in both the statements (1) and (2) together are not sufficient

E. If the data in both the statements (1) and (2) together are necessary

17. How is Rakesh related to Keshav?

(1) Tapan's wife Nisha is paternal aunt of Keshav

(2) Rakesh is the brother of a friend of Nisha

A. if the data in statement I alone are sufficient to answer the question

B. if the data in statement II alone are sufficient answer the question

C. if the data either in I or II alone are sufficient to answer the question

D. if the data even in both the statements together are not sufficient to answer the question

E. If the data in both the statements together are needed

18. In a row of 28 girls and boys how many boys are there to the left of Shridhar?

(1) If Shridhar is placed after two places to the right then he becomes twelfth from the right.

(2) Only every third child in the row is girl

A. If the data in statement (1) alone are sufficient

B. If the data in statement (2) alone are sufficient

C. If the data either in statement (1) alone or in statement (2) alone are sufficient

D. If the data given in both the statements (1) and (2) together are not sufficient

E. If the data in both the statements (1) and (2) together are necessary

19. Who among P, Q, S, T, V and W is the shortest?

(1) S is taller than T, P and W and is not the tallest

(2) T is shorter than Q but is not the shortest

A. If the data in statement (1) alone are sufficient

B. If the data in statement (2) alone are sufficient

C. If the data either in statement (1) alone or in statement (2) alone are sufficient

D. If the data given in both the statements (1) and (2) together are not sufficient

E. If the data in both the statements (1) and (2) together are necessary

20. How is Shubham related to Shivani?

(1) Shubham is brother of Meenal. Shivani is niece of Pooja.

(2) Neeraj is Meenal' s uncle and Preeti' s brother

A. If data in the statement (1) alone is sufficient to answer the question

B. If data in the statement (2) alone is sufficient to answer the question

C. If data either in the statement (1) alone or statement (2) alone are sufficient to answer the question

D. If data given in both (1) & (2) together are not sufficient to answer the question

E. If data in both statements (1) & (2) together are necessary to answer the question

21. In which month of the year was Mohan born?

(1) Mohan was born in winter

(2) Mohan was born exactly fourteen months after his elder sister, who was born in October

A. If the data in statement (1) alone are sufficient

B. If the data in statement (2) alone are sufficient

C. If the data either in statement (1) alone or in statement (2) alone are sufficient

D. If the data given in both the statements (1) and (2) together are not sufficient

E. If the data in both the statements (1) and (2) together are necessary

22. How is PRODUCT written in that code language?

(1) In a certain code language, AIEEE is written as BJFFF.

(2) In a certain code language, GYPSY is written as FXORX

A. If data in the statement (1) alone is sufficient to answer the question

B. If data in the statement (2) alone is sufficient to answer the question

C. If data either in the statement (1) alone or statement (2) alone are sufficient to answer the question

D. If data given in both (1) & (2) together are not sufficient to answer the question

E. If data in both statements (1) & (2) together are necessary to answer the question

IV. Numerical Reasoning (5 questions)

Directions :

Each question is a sequence of numbers with one or two numbers missing. You have to figure out the logical order of the sequence to find out the missing number(s).

23. 5，11，17，25，33，43，( )

A. 49

B. 51

C. 52

D. 53

24. 2，7，19，60，176，( )

A. 530

B. 531

C. 532

D. 533

25. 1，4，7，10，13，( )

A. 14

B. 15

C. 16

D. 17

26. 123，456，789，(　)

A. 1122

B. 101112

C. 11112

D. 100112

27. -9，27，-81，(　)，-729

A. 125

B. -36

C. 360

D. 243

V. Interpretation of Tables and Graphs (8 questions)

Directions :
This is a test on reading and interpretation of data presented in tables and graphs.

Chart 1

Study the following pie-diagrams carefully and answer the questions given below it:

Percentage Composition of Human Body

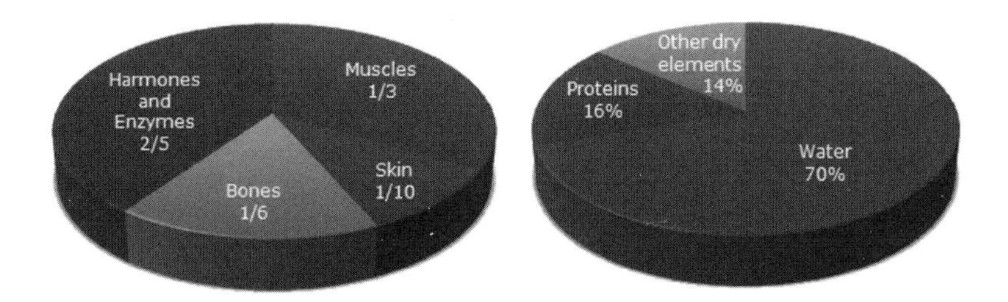

28. What percent of the total weight of human body is equivalent to the weight of the proteins in skin in human body?

    A. 0.016

    B. 1.6

    C. 0.16

    D. Data inadequate

29. What will be the quantity of water in the body of a person weighing 50 kg?

A. 20 kg

B. 35 kg

C. 41 kg

D. 42.5 kg

30. What is the ratio of the distribution of proteins in the muscles to that of the distribution of proteins in the bones?

A. 1:18

B. 1:2

C. 2:1

D. 18:1

31. To show the distribution of proteins and other dry elements in the human body, the arc of the circle should subtend at the centre an angle of:

A. 54°

B. 126°

C. 108°

D. 252°

Chart 2

Study the following line graph which gives the number of students who joined and left the school in the beginning of year for six years, from 1996 to 2001.

Initial Strength of school in 1995 = 3000

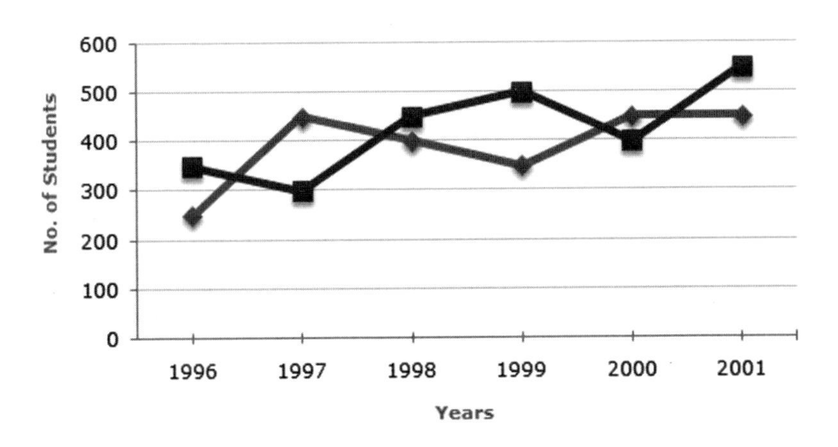

32. The number of students studying in the school during 1999 was?

A. 2950

B. 3000

C. 3100

D. 3150

33. The strength of school increased/decreased from 1997 to 1998 by approximately what percent?

A. 1.2%

B. 1.7%

C. 2.1%

D. 2.4%

34. The number of students studying in the school in 1998 was what percent of the number of students studying in the school in 2001?

A. 92.13%

B. 93.75%

C. 96.88%

D. 97.25%

35. During which of the following pairs of years, the strength of the school was same?

A. 1999 and 2001

B. 1998 and 2000

C. 1997 and 1998

D. 1996 and 2000

全卷完

CRE-APT

文化會社出版社 **CULTURE CROSS LIMITED**

答題紙 ANSWER SHEET

請在此貼上電腦條碼
Please stick the barcode label here

(1) 考生編號 Candidate No.

(2) 考生姓名 Name of Candidate

(3) 考生簽署 Signature of Candidate

宜用H.B.鉛筆作答
You are advised to use H.B. Pencils

考生須依照下圖所示填畫答案：

23 A B C D E

錯填答案可使用潔淨膠擦將筆痕徹底擦去。

切勿摺皺此答題紙

Mark your answer as follows:

23 A B C D E

Wrong marks should be completely erased with a clean rubber.

DO NOT FOLD THIS SHEET

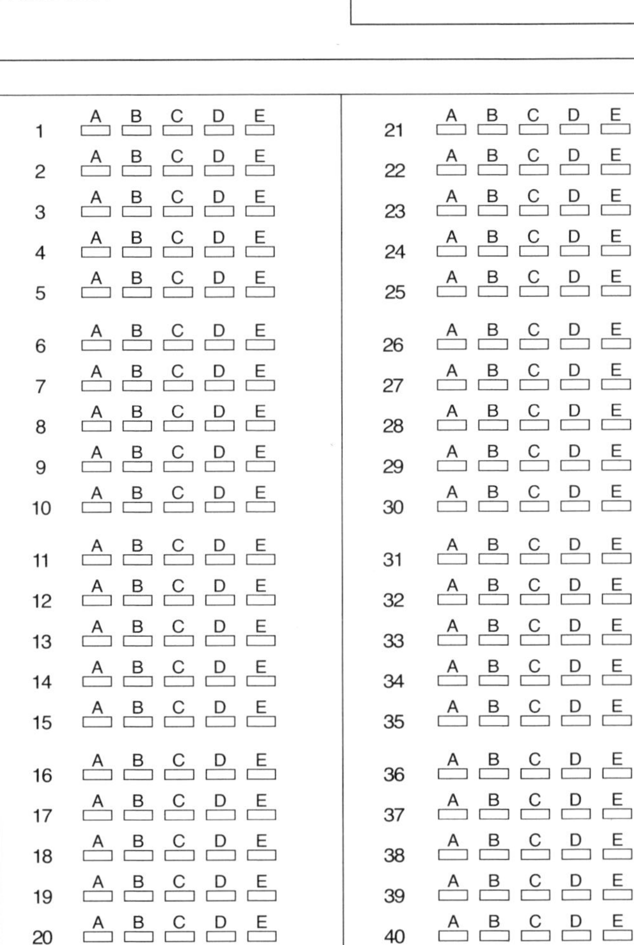

文化會社出版社
投考公務員 模擬試題王

能力傾向試測試
模擬試卷（六）

時間：四十五分鐘

考生須知：

（一）細讀答題紙上的指示。宣布開考後，考生須首先於適當位置貼上電腦條碼及填上各項所需資料。宣布停筆後，考生不會獲得額外時間貼上電腦條碼。

（二）試場主任宣布開卷後，考生請檢查試題冊及確定試題冊內的試題。最後會有「**全卷完**」的字眼。

（三）本試卷各題佔分相等。

（四）**本試卷全部試題均須回答**。為便於修正答案，考生宜用HB鉛筆把答案填畫在答題紙上。錯誤答案可用潔淨膠擦將筆痕徹底擦去。考生須清楚填畫答案，否則會因答案未能被辨認而失分。

（五）每題只可填畫**一個**答案。如填劃超過一個答案，該題將**不獲評分**。

（六）答案錯誤，不另扣分。

（七）未經許多，請勿打開試題冊。

CC-CRE-APT

I. 演繹推理（8題）

請根據以下短文的內容，選出一個或一組推論。請假定短文的內容都是正確的。

1. **某位作家在其晚期的作品中沒有像其早期那樣嚴格遵守小說結構的成規。由於最近發現的一部他的小說的結構像他早期的作品一樣嚴格地遵守了那些成規，因此該作品一定是寫於他的早期。**
   上面論述所依據的假設是：

   A. 該作家在其創作晚期比早期更不願意打破某種成規
   B. 隨著創作的發展，該作家日益意識不到其小說結構的成規
   C. 在其職業生涯晚期，該作家是其時代唯一有意打破小說結構成規的作家
   D. 該作家在其創作生涯的晚期，沒有寫過任何模仿其早期作品風格的小說

2. **儘管新製造的國產汽車的平均油效仍低於新製造的進口汽車，但它在2016年到2019年間卻顯著地提高了。自此以後，新製造的國產汽車的平均油效沒再提高，但新製造的國產汽車與進口汽車平均油效上的差距卻逐漸縮小。**
   如以上論述正確，那麼基於此也一定正確的一項是：

   A. 新製造的進口汽車的平均油效，由2019年後逐漸降低
   B. 新製造的國產汽車的平均油效，由2019年後逐漸降低

C. 2019年後製造的國產汽車的平均油效，高於2019年後製造的進口汽車的平均油效

D. 2016年製造的進口汽車的平均油效，高於2019年製造的進口汽車的平均油效

3. 某珠寶店被盜，警方已發現以下線索：

(1)甲、乙、丙三人中，至少有一人是罪犯

(2)如果甲是罪犯，則乙一定是合謀

(3)盜竊發生時，乙正在咖啡店喝咖啡

可見：

A. 甲是罪犯

B. 甲、乙都是罪犯

C. 甲、乙、丙都是罪犯

D. 丙是罪犯

4. 有A、B、C、D、E五位同學一起比賽象棋，每兩人之間只比賽一盤，統計比賽的盤數知：A賽了4盤，B賽了3盤，C賽了2盤，D賽了1盤，則同學E賽的盤數是：

A. 1盤

B. 2盤

C. 3盤

D. 4盤

5. **城市A是個水資源嚴重缺乏的城市，但長期以來水價格一直偏低。最近市政府根據價值規律擬調高水價，這一舉措將對節約使用該市的水資源產生重大的推動作用。**
若上述結論成立，下列哪些項必須是真的：
(1)有相當數量的水浪費是因為水價格偏低造成的
(2)水價格的上調幅度足以對浪費用水的用戶產生經濟壓力
(3)水價格的上調不會引起用戶的不滿

A. (1)、(2)

B. (1)、(3)

C. (2)、(3)

D. (1)、(2)和(3)

6. **一般病菌多在室溫環境生長繁殖，低溫環境停止生長，僅能維持生命。而耶爾森氏菌卻恰恰相反，不但不怕低溫寒冷，而且只有在0℃左右才大量繁殖。雪櫃裡存儲的食物，使耶爾森氏菌處於生長狀態。**
由此可以推出：

A. 耶爾森氏菌在室溫環境無法生存

B. 一般病菌生長的環境也適合耶爾森氏菌生長

C. 耶爾森氏菌的生長溫度不適合一般病菌

D. 0℃環境下，雪櫃裡僅存在耶爾森氏菌

7. 在同一側的房號為1、2、3、4的四間房裡，分別住著來自韓國、法國、英國和德國的四位專家。有一位記者前來採訪他們，

(1)韓國人説：「我的房號大於德國人，且我不會説外語，也無法和鄰居交流」

(2)法國人説：「我會説德語，但我卻無法和我的鄰居交流」

(3)英國人説：「我會説韓語，但我只可以和一個鄰居交流」

(4)德國人説：「我會説我們這四個國家的語言」

那麼，按照房號按小到大的順序排列，房間裡住的人的國籍依次是：

A. 英國、德國、韓國、法國

B. 法國、英國、德國、韓國

C. 德國、英國、法國、韓國

D. 德國、英國、韓國、法國

8. 未來深海水下線纜的外皮將由玻璃製成，而不是特殊的鋼材或鋁合金。因為金屬具有顆粒狀的微觀結構，在深海壓力之下，粒子交界處的金屬外皮容易斷裂。而玻璃看起來雖然是固體，但在壓力之下可以流動，因此可以視為液體。

由此可以推出：

A. 玻璃沒有顆粒狀的微觀結構

B. 一切固體幾乎都可以被視為緩慢流動的液體

C. 玻璃比起鋼材或鋁合金，更適合做建築材料

D. 與鋼材相比，玻璃的顆粒狀的微觀結構流動性更好

II. Verbal Reasoning (English) (6 questions)

Directions :

In this test, each passage is followed by three statements (the questions). You have to assume what is stated in the passage is true and decide whether the statements are either:

True (Box A): the statement is already made or implied in the passage, or follows logically from the passage.

False (Box B): the statement contradicts what is said, implied by, or follows logically from the passage.

Can't tell (Box C): there is insufficient information in the passage to establish whether the statement is true or false.

Passage 1 (Question 9 to 11):

The United Nations' Convention on International Trade in Endangered Species (CITES. recently reaffirmed a 1989 ban on trading ivory, despite calls from Tanzania and Zambia to lift it. Only 470,000 elephants remain in Africa today – compared to 1.3 million in 1979. While natural habitat loss was a significant factor in dwindling elephant populations, poaching for ivory was the main cause. Since the ban's implementation, elephant populations have recovered in many African countries, but an estimated 38,000 elephants are still killed annually. CITES permitted one-off sales in 1999 and in 2008, allowing approved countries to dispose of their government stockpiles of ivory. Ivory from these sales was exported to Japan and China, where demand for carved ivory is high. Conservation groups vehemently oppose further one-off sales, because much of the ivory sold is of unknown origin. Furthermore, the sales have fuelled far-Eastern demand for ivory. In central and western African countries, where organized crime rings operate lucrative ivory smuggling operations, poaching remains rife. Those in favour of allowing one-off sales argue that elephants are no longer endangered, and that maintaining the ban will simply inflate the price of illegal ivory, making poaching more tempting. Though the CITES decision is viewed as a victory by conservationists, the African elephant' s future relies on governments' commitment to enforcing the ban.

9. No legal sales of Ivory have occurred since 1989.

10. Whether or not African elephants should be classified as endangered is debatable.

11. Conservationists question the provenance of the ivory sold at one-off sales.

Passage 2 (Question 12 to 14):

Globalisation is putting fresh pressure on mid-sized firms, with large companies moving into their markets and dictating prices. This kind of company feels the pressure on all sides. They are too large to qualify for the grants and assistance that small enterprises can count on, but are too small to wield the kind of influence their larger competitors can bring to bear. But despite the squeeze, most mid-sized firms still believe that they can deliver steady growth, by playing to their traditional strength: being more nimble and more customer-focused than their larger rivals.

12. Large companies have a greater effect on prices than mid-sized companies.

13. The size of grant awarded to a company is indirectly proportional to the size of the company.

14. Traditionally mid-sized firms are faster than large companies at adapting to market changes.

III. Data Sufficiency Test (8 questions)

Directions : In this test, you are required to choose a combination of clues to solve a problem.

15. How many employees of bank 'X' opted for VRS?

    (1) 18% of the 950 officer cadre employees and 6% of the 1100 of all other cadre employees opted for VRS.

    (2) 28% of the employees in the age group of 51 to 56 and 17% of the employees in all other age groups opted for VRS.

    A. If the data in statement (1) alone are sufficient

    B. If the data in statement (2) alone are sufficient

    C. If the data either in statement (1) alone or in statement (2) alone are sufficient

    D. If the data given in both the statements (1) and (2) together are not sufficient

    E. If the data in both the statements (1) and (2) together are necessary

16. What time did the train leave today?

(1) The train normally leaves on time

(2) The scheduled departure is at 14:30

A. if the data in statement (1) alone are sufficient to answer the question

B. if the data in statement (2) alone are sufficient answer the question

C. if the data either in (1) or (2) alone are sufficient to answer the question

D. if the data even in both the statements together are not sufficient to answer the question

E. If the data in both the statements together are needed

17. What is Sunil’s position in a row of forty students?

(1) There are 16 students towards the left of Sunil

(2) There are 23 students towards the right of Sunil

A. If the data in statement (1) alone are sufficient

B. If the data in statement (2) alone are sufficient

C. If the data either in statement (1) alone or in statement (2) alone are sufficient

D. If the data given in both the statements (1) and (2) together are not sufficient

E. If the data in both the statements (1) and (2) together are necessary

18. The annual function of college 'X' was celebrated on which date?

(1) The annual function was celebrated on its 25th foundation day

(2) The college was founded on 85th day of the year 1974

A. If the data in statement (1) alone are sufficient

B. If the data in statement (2) alone are sufficient

C. If the data either in statement (1) alone or in statement (2) alone are sufficient

D. If the data given in both the statements (1) and (2) together are not sufficient

E. If the data in both the statements (1) and (2) together are necessary

19. Who is taller among P, Q, R, S & T?

(1) S is shorter than Q. P is shorter than only T

(2) Q is taller than only S. T is taller than P and R

A. If data in the statement (1) alone is sufficient to answer the question

B. If data in the statement (2) alone is sufficient to answer the question

C. If data either in the statement (1) alone or statement (2) alone are sufficient to answer the question

D. If data given in both (1) & (2) together are not sufficient to answer the question

E. If data in both statements (1) & (2) together are necessary to answer the question

20. “You must submit your application with in 10 days from, the date of release of this advertisement.” What is the exact date before which the application must be submitted?

(1) The advertisement was released on 18th February

(2) It was a leap year

A. If the data in statement (1) alone are sufficient

B. If the data in statement (2) alone are sufficient

C. If the data either in statement (1) alone or in statement (2) alone are sufficient

D. If the data given in both the statements (1) and (2) together are not sufficient

E. If the data in both the statements (1) and (2) together are necessary

21. What is the distance between point P and point Q?

(1) Point R is 10 m west of point P and point S is 10 m north of point P

(2) Point Q is 10 m south-east of point R. Point S is 20 m north-west of point Q

A. If data in the statement (1) alone is sufficient to answer the question

B. If data in the statement (2) alone is sufficient to answer the question

C. If data either in the statement (1) alone or statement (2) alone are sufficient to answer the question

D. If data given in both (1) & (2) together are not sufficient to answer the question

E. If data in both statements (1) & (2) together are necessary to answer the question

22. What is the colour of the fresh grass?

(1) Blue is called green, red is called orange, orange is called yellow

(2) Yellow is called white, white is called black, green is called brown and brown is called purple

A. If the data in statement (1) alone are sufficient

B. If the data in statement (2) alone are sufficient

C. If the data either in statement (1) alone or in statement (2) alone are sufficient

D. If the data given in both the statements (1) and (2) together are not sufficient

E. If the data in both the statements (1) and (2) together are necessary

IV. Numerical Reasoning (5 questions)

Directions :

Each question is a sequence of numbers with one or two numbers missing. You have to figure out the logical order of the sequence to find out the missing number(s).

23. 8/9，-2/3，1/2，-3/8，( )

 A. 9/32

 B. 5/72

 C. 8/32

 D. 9/23

24. 9，35，79，141，221，( )

 A. 357

 B. 319

 C. 303

 D. 251

25. 8，48，120，224，360，( )

 A. 528

 B. 562

 C. 626

 D. 682

26. 1，10，31，70，133，( )

A. 136

B. 186

C. 226

D. 256

27. -2，1，31，70，112，( )

A. 154

B. 155

C. 256

D. 280

V. Interpretation of Tables and Graphs (8 questions)

Directions :
This is a test on reading and interpretation of data presented in tables and graphs.

Study the following table and answer the questions.

Chart 1

Classification of 100 Students Based on the Marks Obtained by them in Physics and Chemistry in an Examination.

Subject	Marks out of 50				
	40 and above	30 and above	20 and above	10 and above	0 and above
Physics	9	32	80	92	100
Chemistry	4	21	66	81	100
Average (Aggregate)	7	27	73	87	100

28. What is the different between the number of students passed with 30 as cut-off marks in Chemistry and those passed with 30 as cut-off marks in aggregate?

A. 3

B. 4

C. 5

D. 6

29. If at least 60% marks in Physics are required for pursuing higher studies in Physics, how many students will be eligible to pursue higher studies in Physics?

A. 27

B. 32

C. 34

D. 41

30. The percentage of number of students getting at least 60% marks in Chemistry over those getting at least 40% marks in aggregate, is approximately?

A. 21%

B. 27%

C. 29%

D. 31%

31. The number of students scoring less than 40% marks in aggregate is?

A. 13

B. 19

C. 20

D. 27

Chart 2

Study the bar chart and answer the question based on it.

Production of Fertilizers by a Company (in 1000 tonnes) Over the Years

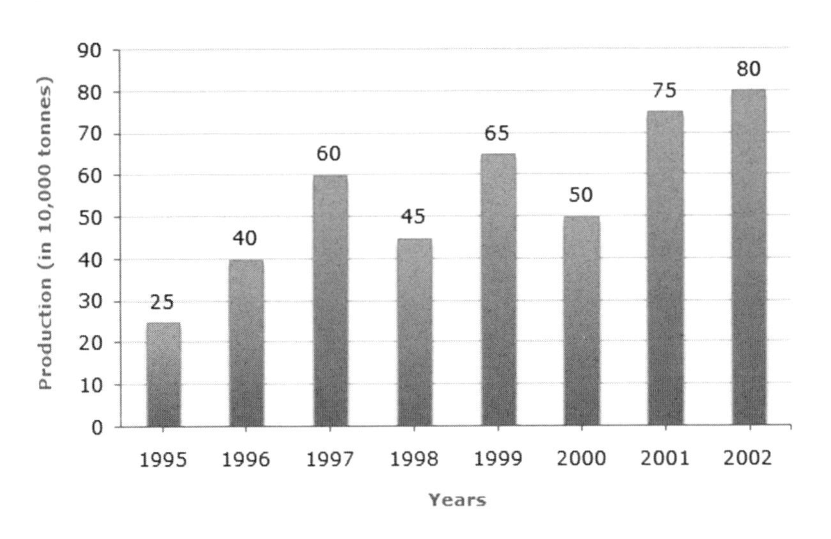

32. What was the percentage decline in the production of fertilizers from 1997 to 1998?

    A. 33(1/3)%

    B. 20%

    C. 25%

    D. 21%

33. The average production of 1996 and 1997 was exactly equal to the average production of which of the following pairs of years?

A. 2000 and 2001

B. 1999 and 2000

C. 1998 and 2000

D. 1995 and 2001

34. What was the percentage increase in production of fertilizers in 2002 compared to that in 1995?

A. 320%

B. 300%

C. 220%

D. 200%

35. In which year was the percentage increase in production as compared to the previous year the maximum?

A. 2002

B. 2001

C. 1997

D. 1996

全卷完

CRE-APT

文化會社出版社 **CULTURE CROSS LIMITED**

答題紙 ANSWER SHEET

請在此貼上電腦條碼 Please stick the barcode label here

(1) 考生編號 Candidate No.

(2) 考生姓名 Name of Candidate

(3) 考生簽署 Signature of Candidate

宜用H.B.鉛筆作答
You are advised to use H.B. Pencils

考生須依照下圖所示填畫答案：

23 A B C D E

錯填答案可使用潔淨膠擦將筆痕徹底擦去。

切勿摺皺此答題紙

Mark your answer as follows:

23 A B C D E

Wrong marks should be completely erased with a clean rubber.

DO NOT FOLD THIS SHEET

No.						No.					
1	A	B	C	D	E	21	A	B	C	D	E
2	A	B	C	D	E	22	A	B	C	D	E
3	A	B	C	D	E	23	A	B	C	D	E
4	A	B	C	D	E	24	A	B	C	D	E
5	A	B	C	D	E	25	A	B	C	D	E
6	A	B	C	D	E	26	A	B	C	D	E
7	A	B	C	D	E	27	A	B	C	D	E
8	A	B	C	D	E	28	A	B	C	D	E
9	A	B	C	D	E	29	A	B	C	D	E
10	A	B	C	D	E	30	A	B	C	D	E
11	A	B	C	D	E	31	A	B	C	D	E
12	A	B	C	D	E	32	A	B	C	D	E
13	A	B	C	D	E	33	A	B	C	D	E
14	A	B	C	D	E	34	A	B	C	D	E
15	A	B	C	D	E	35	A	B	C	D	E
16	A	B	C	D	E	36	A	B	C	D	E
17	A	B	C	D	E	37	A	B	C	D	E
18	A	B	C	D	E	38	A	B	C	D	E
19	A	B	C	D	E	39	A	B	C	D	E
20	A	B	C	D	E	40	A	B	C	D	E

文化會社出版社
投考公務員 模擬試題王

能力傾向試測試
模擬試卷（七）

時間：四十五分鐘

考生須知：

(一) 細讀答題紙上的指示。宣布開考後，考生須首先於適當位置貼上電腦條碼及填上各項所需資料。宣布停筆後，考生不會獲得額外時間貼上電腦條碼。

(二) 試場主任宣布開卷後，考生請檢查試題冊及確定試題冊內的試題。最後會有「**全卷完**」的字眼。

(三) 本試卷各題佔分相等。

(四) **本試卷全部試題均須回答**。為便於修正答案，考生宜用HB鉛筆把答案填畫在答題紙上。錯誤答案可用潔淨膠擦將筆痕徹底擦去。考生須清楚填畫答案，否則會因答案未能被辨認而失分。

(五) 每題只可填畫**一個**答案。如填劃超過一個答案，該題將**不獲評分**。

(六) 答案錯誤，不另扣分。

(七) 未經許多，請勿打開試題冊。

CC-CRE-APT

I. 演繹推理（8題）

請根據以下短文的內容，選出一個或一組推論。請假定短文的內容都是正確的。

1. **糧食可以在收割前在期貨市場進行交易。如果預測水稻產量不足，水稻期貨價格就會上升；如果預測水稻豐收，水稻期貨價格就會下降。假設今天早上，氣象學家們預測從明天開始水稻產區會有適量降雨。因為充分的潮濕對目前水稻的生長非常重要，所以今天的水稻期貨價格會大幅下降。**
   下面哪項如果正確，最嚴重地削弱以上的觀點？

   A. 農業專家們今天宣布，一種水稻病菌正在傳播

   B. 本季度水稻期貨價格的波動比上季度更加劇烈

   C. 氣象學家們預測的明天的降雨估計很可能會延伸到穀物產區以外

   D. 在關鍵的授粉階段沒有接受足夠潮濕的穀物不會取得豐收

2. **熱天會使人煩躁不安，對他人採取負面反應，甚至進攻，發生反社會行為。世界上炎熱的地方，也是攻擊行為較多的地方。由此可推出：**

   A. 自然環境決定人格特徵

   B. 自然物理環境可對特定行為做出一定的解釋

   C. 越是寒冷的地方，人們越不會出現侵犯行為

   D. 炎熱的地方社會治安更好

3. 很久以前，在法國薯仔被稱為「鬼蘋果」，農民們都不願意引種。一位農學家想出一個方法，在一塊土地上種植薯仔，並由一支著軍禮服、全副武裝的國王衛隊看守，到了夜晚，衛隊故意撤走。結果人們紛紛來偷薯仔，引種到自己田裡，通過這種方法，薯仔的種籽在法國得到迅速的推廣，由此可推出：

A. 有些東西越禁止，就越引起人們的興趣，比如某些電影、書籍越禁止越走俏

B. 人們都有獵奇心理

C. 人們都有違反規定、打破限制的傾向

D. 新事物的出現，開始都是不受歡迎的

4. 1964年美國紐約發生了著名的吉諾維斯案件，一位叫做吉諾維斯的姑娘在回家途中遭歹徒持刀殺害。案發的30分鐘內有38個鄰居聽到被害者的呼救聲，許多人還走到窗前看了很長時間，但沒有一個人去救援。甚至沒有人行舉手之勞，打電話及時報警。致使一件不該發生的慘劇成為現實。
對於上述現象的解釋最恰當的一項是：

A. 公眾目睹別人身臨危難，是公眾人性的喪失

B. 經常目睹暴力事件，導致公眾的麻木不仁

C. 經常目睹事件發生，而不予幫助，可能是由於其他群眾在場，抑制了助人動機，導致社會責任的分散

D. 面對危險人們都有自保的傾向

5. **研究表明，美國和德國的青少年將父母的「限制管教」的方式看作是討厭自己，而在韓國和日本，父母同樣的管教方式卻使孩子感受到的是接納和溫暖。**
由此可推出：

A. 韓、日兒童更容易被管教

B. 美、德的孩子比較起韓、日兩國的孩子，更具有逆反心理

C. 美、德兩國更注重人性化的教育方式

D. 東西方文化背景，影響到兒童的教育方式

6. **在城市一條大街上，一家百貨商店被人盜竊了一批財物。事情發生後，警察經過偵察拘捕了3個嫌疑犯：甲、乙、丙。後來，又經審訊，查明了以下事實：**
(1)罪犯帶著贓物是開車逃掉的
(2)假如不伙同甲，丙決不會犯案
(3)乙不會開車
(4)罪犯就是這三個人中的一個或一伙
由此一定可推出：

A. 甲有罪

B. 甲無罪

C. 乙有罪

D. 乙無罪

7. **野生大熊貓正在迅速減少。因此，為了保護該物種，應把現存的野生大熊貓捕捉起來，並放到世界名地的動物園裡去。**
以下哪項，如果正確，對上述結論提出了最嚴重的質疑？

A. 野生大熊貓在關起來時通常會比在野生棲身地時生下更多的仔

B. 在動物園中剛生下來的大熊貓不容易死於傳染病，但是野生大熊貓很可能死於這些疾病

C. 在野生大熊貓的棲息地以外，很難弄到足夠數量的竹，這是大熊貓唯一的食物

D. 動物園裡的大熊貓和野生的大熊貓後代中能夠活到成年的個體數量相當

8. **對許多科學家來說，克隆技術為更有效地設計新的生命形式、拯救瀕危物種以及探索多種人類疾病的治療方法提供了前所未有的機會。**
由此可推出：

A. 克隆技術將會推動科學和社會的進步

B. 科學家們都認為需要進一步發展克隆技術

C. 許多科學家將會反對政府對研製克隆技術的壓制

D. 隨著克隆技術的發展，克隆人的出現將不可避免

II. Verbal Reasoning (English) (6 questions)

Directions :

In this test, each passage is followed by three statements (the questions). You have to assume what is stated in the passage is true and decide whether the statements are either:

True (Box A): the statement is already made or implied in the passage, or follows logically from the passage.

False (Box B): the statement contradicts what is said, implied by, or follows logically from the passage.

Can' t tell (Box C): there is insufficient information in the passage to establish whether the statement is true or false.

Passage 1 (Question 9 to 11):

If society seems obsessed with youth, it is at least partly because companies are. Like it or not, the young increasingly pick the styles and brands that trickle up to the rest of the population. Nike, Abercrombie & Fitch and Timberland first found success with the young, and when that clientele tired of them the companies felt the loss deeply. Now that adults are no longer necessarily expected to act and look grown-up, parents and children can be found listening to exactly the same music, playing the same computer games, watching the same TV programmes, and wearing the same brands of clothes and shoes.

9. An adult' s style can sometimes be similar to that of a child' s.

10. The profits of Timberland are not affected by young customers.

11. Adults wear the same shoes as children because they want to look younger.

Passage 2 (Question 12 to 14):

Television is changing as it goes digital. The result will not only be better-quality pictures and sound but also personal TV, with viewers able to tailor the programmes they watch and even interact with them. How much money this will make for programme producers or broadcasters, whoever they may be, is not so clear.

Cable, satellite and terrestrial television broadcasters are upgrading their equipment to provide higher quality digital services. Rupert Murdoch's News Corporation will become the first company in the world to migrate an entire national TV system over to digital when it turns off its old analogue version of its British satellite service, BSkyB.

12. Rupert Murdoch is associated with BSkyB.

13. The only change from traditional analogue services to digital services will be the picture quality.

14. Television broadcasters are upgrading their equipment because they will make more money from digital TV.

III. Data Sufficiency Test (8 questions)

Directions : In this test, you are required to choose a combination of clues to solve a problem.

15. Is D brother of F?

    (1) B has two sons of which F is one

    (2) D' s mother is married to B

    A. if the data in statement (1) alone are sufficient to answer the question

    B. if the data in statement (2) alone are sufficient answer the question

    C. if the data either in (1) or (2) alone are sufficient to answer the question

    D. if the data even in both the statements together are not sufficient to answer the question

    E. If the data in both the statements together are needed

16. Five friends P, Q, R, S and T are standing in a row facing East. Who is standing at the extreme right end?

    (1) Only P is between S and T; R is to the immediate right of T

    (2) R is between T and Q

    A. If the data in statement (1) alone are sufficient

    B. If the data in statement (2) alone are sufficient

    C. If the data either in statement (1) alone or in statement (2) alone are sufficient

D. If the data given in both the statements I and II together are not sufficient

E. If the data in both the statements I and II together are necessary

17. Among four friends A, B, C and D, who is the heaviest?

(1) B is heavier than A, but lighter than D

(2) C is lighter than B

A. if the data in statement (1) alone are sufficient to answer the question

B. if the data in statement (2) alone are sufficient answer the question

C. if the data either in (1) or (2) alone are sufficient to answer the question

D. if the data even in both the statements together are not sufficient to answer the question

E. if the data in both the statements together are needed.

18. In which month did Rahul go to Kanpur for business?

(1) Rahul' s son remembers that he went after 20th August but before 10th September

(2) Varun, friend of Rahul remembers that he went Kanpur in the 3rd quarter of the fiscal year

A. If data in the statement (1) alone is sufficient to answer the question

B. If data in the statement (2) alone is sufficient to answer the question

C. If data either in the statement (1) alone or statement (2) alone are sufficient to answer the question

D. If data given in both (1) & (2) together are not sufficient to answer the question

E. If data in both statements (1) & (2) together are necessary to answer the question

19. Which code word stands for 'good' in the coded sentence 'sin co bye' which means 'He is good' ?

(1) In the same code language 'co mot det' means 'They are good'

(2) In the same code language 'sin mic bye' means 'He is honest'

A. If the data in statement (1) alone are sufficient

B. If the data in statement (2) alone are sufficient

C. If the data either in statement (1) alone or in statement (2) alone are sufficient

D. If the data given in both the statements (1) and (2) together are not sufficient

E. If the data in both the statements (1) and (2) together are necessary

20. Madan' s flat is on which floor of 5 floor apartments?

(1) Harish flat, which is adjacent to Madan, is exactly below Karan' s flat which is on fifth floor

(2) Madan' s flat is exactly above Gopal' s flat, whose flat is exactly above Nitin' s first floor flat

A. If data in the statement (1) alone is sufficient to answer the question

B. If data in the statement (2) alone is sufficient to answer the question

C. If data either in the statement (1) alone or statement (2) alone are sufficient to answer the question

D. If data given in both (1) & (2) together are not sufficient to answer the question

E. If data in both statements (1) & (2) together are necessary to answer the question

21. How many pencils does the shopkeeper sells on Sunday?

(1) On Sunday he sold 12 more pencils than he sold the previous day

(2) He sold 28 pencils each on Thursday and Saturday

A. If data in the statement (1) alone is sufficient to answer the question

B. If data in the statement (2) alone is sufficient to answer the question

C. If data either in the statement (1) alone or statement (2) alone are sufficient to answer the question

D. If data given in both (1) & (2) together are not sufficient to answer the question

E. If data in both statements (1) & (2) together are necessary to answer the question

22. What is the code for 'is' in the code language?

(1) In the code language, 'shi tu ke' means 'pen is blue'

(2) In the same code language, 'ke si re' means 'this is wonderful'

A. If the data in statement (1) alone are sufficient

B. If the data in statement (2) alone are sufficient

C. If the data either in statement (1) alone or in statement (2) alone are sufficient

D. If the data given in both the statements (1) and (2) together are not sufficient

E. If the data in both the statements (1) and (2) together are necessary

IV. Numerical Reasoning (5 questions)

Directions :
Each question is a sequence of numbers with one or two numbers missing. You have to figure out the logical order of the sequence to find out the missing number(s).

23. 1，1，3，5，11，( )

A. 8

B. 13

C. 21

D. 32

24. 102，1030204，10305020406，( )

A. 10305070206

B. 103054008

C. 103050702040608

D. 10305608

25. 1，2，2，4，4，6，8，8，( )

A. 12

B. 14

C. 10

D. 16

26. 2，6，18，54，(　)

A. 186

B. 162

C. 194

D. 196

27. 6/28，21/98，18/84，9/42，(　)

A. 25/60

B. 12/44

C. 12/56

D. 25/78

V. Interpretation of Tables and Graphs (8 questions)

Directions :
This is a test on reading and interpretation of data presented in tables and graphs.

Chart 1

Study the following graph and the table and answer the questions given below.

Data of different states regarding population of states in the year 1998

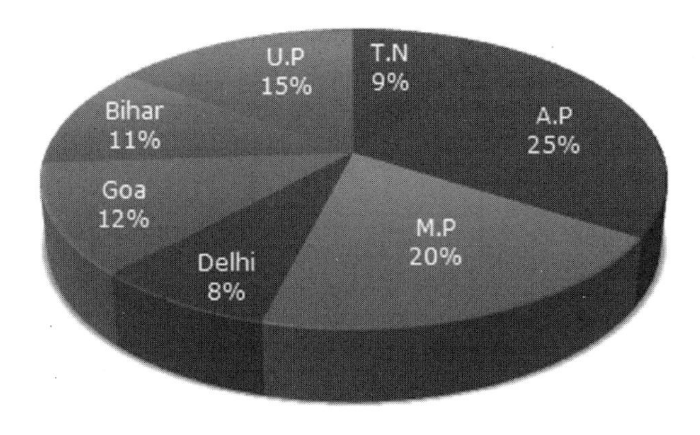

Total population of the given States = 3276000

States	Sex and Literacy wise Population Ratio			
	Sex		Literacy	
	M	F	Literate	Illiterate
A.P	5	3	2	7
M.P	3	1	1	4
Delhi	2	3	2	1
Goa	3	5	3	2
Bihar	3	4	4	1
U.P.	3	2	7	2
T.N.	3	4	9	4

28. What will be the percentage of total number of males in U.P., M.P. and Goa together to the total population of all the given states?

A. 25%

B. 27.5%

C. 28.5%

D. 31.5%

29. What was the total number of illiterate people in A.P. and M.P. in 1998?

A. 876040

B. 932170

C. 981550

D. 1161160

30. What is the ratio of the number of females in T.N. to the number of females in Delhi?

A. 7:5

B. 9:7

C. 13:11

D. 15:14

31. What was the number of males in U.P. in the year 1998?

A. 254650

B. 294840

C. 321470

D. 341200

Chart 2

In a school the periodical examination are held every second month. In a session during April 2001 - March 2002, a student of Class IX appeared for each of the periodical exams. The aggregate marks obtained by him in each perodical exam are represented in the line-graph given below.

Marks Obtained by student in Six Periodical Held in Every Two Months During the Year in the Session 2001 - 2002.

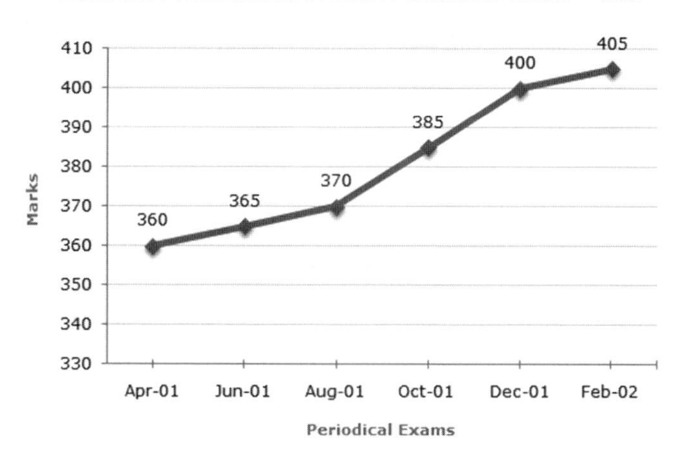

32. In which periodical exams did the student obtain the highest percentage increase in marks over the previous periodical exams ?

A. June, 01

B. August, 01

C. Oct, 01

D. Dec, 01

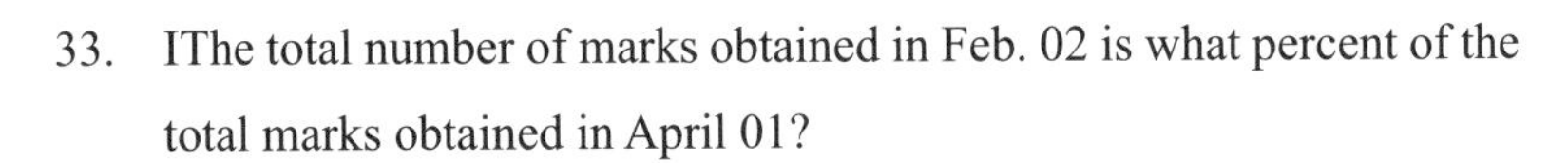

33. IThe total number of marks obtained in Feb. 02 is what percent of the total marks obtained in April 01?

A. 110%

B. 112.5%

C. 115%

D. 116.5%

34. What is the percentage of marks obtained by the student in the periodical exams of August, 01 and Oct, 01 taken together?

A. 73.25%

B. 75.5%

C. 77%

D. 78.75%

35. What are the average marks obtained by the student in all the periodical exams during the last session?

A. 373

B. 379

C. 381

D. 385

全卷完

CRE-APT

文化會社出版社 **CULTURE CROSS LIMITED**

答題紙 ANSWER SHEET

請在此貼上電腦條碼
Please stick the barcode label here

(1) 考生編號 Candidate No.

(2) 考生姓名 Name of Candidate

(3) 考生簽署 Signature of Candidate

宜用H.B.鉛筆作答
You are advised to use H.B. Pencils

考生須依照下圖所示填畫答案：

23 A B C D E

錯填答案可使用潔淨膠擦將筆痕徹底擦去。

切勿摺皺此答題紙

Mark your answer as follows:

23 A B C D E

Wrong marks should be completely erased with a clean rubber.

DO NOT FOLD THIS SHEET

1	A	B	C	D	E	21	A	B	C	D	E
2	A	B	C	D	E	22	A	B	C	D	E
3	A	B	C	D	E	23	A	B	C	D	E
4	A	B	C	D	E	24	A	B	C	D	E
5	A	B	C	D	E	25	A	B	C	D	E
6	A	B	C	D	E	26	A	B	C	D	E
7	A	B	C	D	E	27	A	B	C	D	E
8	A	B	C	D	E	28	A	B	C	D	E
9	A	B	C	D	E	29	A	B	C	D	E
10	A	B	C	D	E	30	A	B	C	D	E
11	A	B	C	D	E	31	A	B	C	D	E
12	A	B	C	D	E	32	A	B	C	D	E
13	A	B	C	D	E	33	A	B	C	D	E
14	A	B	C	D	E	34	A	B	C	D	E
15	A	B	C	D	E	35	A	B	C	D	E
16	A	B	C	D	E	36	A	B	C	D	E
17	A	B	C	D	E	37	A	B	C	D	E
18	A	B	C	D	E	38	A	B	C	D	E
19	A	B	C	D	E	39	A	B	C	D	E
20	A	B	C	D	E	40	A	B	C	D	E

文化會社出版社
投考公務員 模擬試題王

能力傾向試測試
模擬試卷（八）

時間：四十五分鐘

考生須知：

（一）細讀答題紙上的指示。宣布開考後，考生須首先於適當位置貼上電腦條碼及填上各項所需資料。宣布停筆後，考生不會獲得額外時間貼上電腦條碼。

（二）試場主任宣布開卷後，考生請檢查試題冊及確定試題冊內的試題。最後會有「**全卷完**」的字眼。

（三）本試卷各題佔分相等。

（四）**本試卷全部試題均須回答**。為便於修正答案，考生宜用HB鉛筆把答案填畫在答題紙上。錯誤答案可用潔淨膠擦將筆痕徹底擦去。考生須清楚填畫答案，否則會因答案未能被辨認而失分。

（五）每題只可填畫**一個**答案。如填劃超過一個答案，該題將**不獲評分**。

（六）答案錯誤，不另扣分。

（七）未經許多，請勿打開試題冊。

CC-CRE-APT

I. 演繹推理（8題）

請根據以下短文的內容，選出一個或一組推論。請假定短文的內容都是正確的。

1. **不管你使用哪種牙膏，經常刷牙將降低你牙齒腐爛的可能性，科學家得出結論：當刷牙時，通過去除牙齒與牙齦上所形成的牙菌斑薄片，可降低牙齒腐爛的可能性。因此你可以不用加入「氟」的牙膏，只要認真刷牙就能告別蛀牙。**

   下面哪一項是對上面論述推理的一個批評？

   A. 用加入氟的牙膏刷牙，可以降低牙齒腐爛的可能性

   B. 刷牙可降低牙齒腐爛可能性的事實，並沒有表明「氟」不起任何作用

   C. 幾乎沒有人可以通過刷牙，充分地消除牙菌斑

   D. 在絕大多數的時間內，人們的牙齒上都有牙菌斑

2. **有位外國遊客說：「不到鼓山，不算遊福州；不看磨崖石刻，白來鼓山。」**

   根據這句話，最有可能推出的結論是：

   A. 遊福州，只要看磨崖石刻就可以了

   B. 鼓山的磨崖石刻集中了福州最精彩的人文景觀

   C. 遊福州，最令人難忘的是磨崖石刻

   D. 遊福州，應先看磨崖石刻

3. **甲先生、乙先生、丙先生三人是同一家公司的員工，他們的未婚妻A女士、B女士、C女士也都是該公司的職員。知情者介紹說：「A女士的未婚夫是乙先生的好友，並在三個男子中最年輕；丙先生的年齡比C女士的未婚夫大。」**
依據該知情者提供的情況，我們可以推出三對組合分別是：

A. 甲先生—A女士，乙先生—B女士，丙先生—C女士

B. 甲先生—B女士，乙先生—A女士，丙先生—C女士

C. 甲先生—C女士，乙先生—B女士，丙先生—A女士

D. 甲先生—A女士，乙先生—C女士，丙先生—B女士

4. **如果你的手提電腦是2018年以後製造的，那麼它就帶有調製解調器。**
以上斷定可由哪個選項推出？

A. 只有2018年以後製造的筆記本計算機才帶有調制解調器

B. 所有2018年以後製造的筆記本計算機都帶有調制解調器

C. 有些2018年以後製造的筆記本計算機也帶有調制解調器

D. 所有2018年以後製造的筆記本計算機都不帶有調制解調器

5. 有的地質學家認為，如果地球的未勘探地區中單位面積的平均石油儲藏量能和已勘探地區一樣的話，那麼，目前關於地下未開採的能源含量的正確估計因此要乘上一萬倍，如果地質學家的這一觀點成立，那麼，我們可以得出結論：地球上未勘探地區的總面積是已勘探地區的一萬倍。

為使上述論證成立，以下哪些是必須假設的？

(1)目前關於地下未開採的能源含量的估計，只限於對已勘探地區

(2)目前關於地下未開採的能源含量的估計，只限於對石油含量

(3)未勘探地區中的石油儲藏能和已勘探地區一樣得到有效的勘探和開採

A. 只有(1)

B. 只有(2)

C. 只有(1)和(2)

D. (1)、(2)、(3)

6. 在一項實驗中，實驗對象的一半作為「實驗組」，食用了大量的某種辣椒。而作為「對照組」的另一半沒有吃這種辣椒。結果，實驗組的認知能力比對照組差得多。這一結果是由於這種辣椒的一種主要成分——維生素E造成的。

以下哪項如果為真，則最有助於證明這種辣椒中成分造成這一實驗結論？

A. 上述結論中所提到的維生素E在所有蔬菜中都有，為了保證營養必須攝入一定份量這種維生素E

B. 實驗組中人們所食用的辣椒數量，是在政府食品條例規定的安全用量之內的

C. 第二次實驗時，只給一組食用大量辣椒作為實驗組，而不高於不食用辣椒的對照組

D. 實驗前兩組實驗對象，是按認知能力作出「均等劃分」

7. **乒乓球教練將從右手執拍的選手A、B、C和左手執拍的W、Y、Z中選出四名隊員參加奧運會，要求至少有兩名右手執拍的選手，而且選出的四名隊員都可以互相配對進行對打，已知B不能與W配對，C不能與Y配對，X不能與W或Y配對，若A不能入選，那麼有幾種選法：**

A. 一

B. 二

C. 三

D. 四

8. **某家庭有6個孩子，3個孩子是女孩。其中5個孩子有雀斑，4個孩子鬈髮。**

這表明有可能：

A. 兩個男孩鬈髮但沒有雀斑

B. 三個有雀斑的女孩都沒有鬈髮

C. 兩個有雀斑的男孩都沒有鬈髮

D. 三個有鬈髮的男孩只有一個有雀斑

II. Verbal Reasoning (English) (6 questions)

Directions :
In this test, each passage is followed by three statements (the questions). You have to assume what is stated in the passage is true and decide whether the statements are either:
True (Box A): the statement is already made or implied in the passage, or follows logically from the passage.
False (Box B): the statement contradicts what is said, implied by, or follows logically from the passage.
Can't tell (Box C): there is insufficient information in the passage to establish whether the statement is true or false.

Passage 1 (Question 9 to 11):

There are 562 federally recognized American Indian tribes, with a total of
1.7 million members. Additionally, there are hundreds of groups seeking federal recognition – or sovereignty – though less than ten percent will successfully achieve this status. Federally recognised tribes have the right to self-government, and are also eligible for federal assistance programmes. Exempt from state and local jurisdiction, tribes may enforce their own laws, request tax breaks and control regulatory activities. There are however limitations to their sovereignty including, amongst others, the ability to make war and create currency. Historically, tribes were granted

federal recognition through treaties or by executive order. Since 1978 however, this has been replaced by a lengthy and stringent regulatory process which requires tribes applying for federal recognition to fulfil seven criteria, such as anthropological and historical evidence. One of the complications regarding federal recognition is the legal definition of "Indian". Previously, racial criteria, tribal records and personal affidavits were used to classify American Indians. Since the 1970s, however, there has been a shift to the use of a political definition – requiring membership in a federally recognized tribe in order to qualify for benefits, such as loans and educational grants. This definition, however, excludes many individuals of Native American heritage who are not tribal members.

9. There are only two exemptions to a federeally recogized tribe' s powers of self-government.

10. Demand for federal recognition is high because it is a prerequisite for benefit programmes.

11. Since 1978 it has become harder for a tribe to achieve federally recognized status.

Passage 2 (Question 12 to 14):

The first problem with financial statements is that they are in the past; however detailed, they provide just a snap-shot of the business at one moment in time. There is also a lack of detail in financial statements, giving little use in the running of a business. Financial statements are provided for legal reasons to meet with accounting regulations and are used mainly by City analysts who compute share prices and give guidance to shareholders. Accounts often have hidden information and may also be inconsistent; it is difficult to compare different companies' accounts, despite there being standards, as there is much leeway in the standards.

12. Financial statements are useful for businesses to understand their financial activities.

13. Companies create financial statements in order to comply with their legal obligations.

14. If account reporting standards were tightened, it would be easier to compare the performance of different companies.

III. Data Sufficiency Test (8 questions)

Directions : In this test, you are required to choose a combination of clues to solve a problem.

15. How many students are there between Ram and Aditya a row of 40 students?

    (1) Ram is 7th from the left end, Aditya is 17th from the right end.

    (2) Ram is 6 places away from Jeni, who is 20th from the left end.

    A. If data in the statement (1) alone is sufficient to answer the question.

    B. If data in the statement (2) alone is sufficient to answer the question.

    C. If data either in the statement (1) alone or statement (2) alone are sufficient to answer the question.

    D. If data given in both (1) & (2) together are not sufficient to answer the question.

    E. If data in both statements (1) & (2) together are necessary to answer the question.

16. How is A related to C?

    (1) A is the son of C' s grandfather B

    (2) The sister of C is mother of A' s son D

    A. If data in the statement (1) alone is sufficient to answer the question.

    B. If data in the statement (2) alone is sufficient to answer the question.

C. If data either in the statement (1) alone or statement (2) alone are sufficient to answer the question.

D. If data given in both (1) & (2) together are not sufficient to answer the question.

E. If data in both statements (1) & (2) together are necessary to answer the question.

17. When is Manohar's birthday this year?

(1) It is between January 13 and 15, January 13 being Wednesday.

(2) It is not on Friday.

A. if the data in statement (1) alone are sufficient to answer the question

B. if the data in statement (2) alone are sufficient answer the question

C. if the data either in (1) or (2) alone are sufficient to answer the question

D. if the data even in both the statements together are not sufficient to answer the question

E. if the data in both the statements together are needed

18. When was Samir born?

(1) Samir passed out from the university on his 22nd birth-day on 16th April 1999.

(2) Samir was elder than Sudha by three years who recently celebrated her 18th birthday.

A. If the data in statement (1) alone are sufficient

B. If the data in statement (2) alone are sufficient

C. If the data either in statement (1) alone or in statement (2) alone are sufficient

D. If the data given in both the statements (1) and (2) together are not sufficient

E. If the data in both the statements (1) and (2) together are necessary

19. How is GREEN written in a code language?

(1) GREEN AND BLACK is coded as ‘ #@7’ and ORANGE AND PINK is coded as ‘$%#’

(2) PINK AND RED is coded as ‘ #$8’ and YELLOW AND GREEN is coded as ‘6@#’

A. If data in the statement (1) alone is sufficient to answer the question

B. If data in the statement (2) alone is sufficient to answer the question

C. If data either in the statement (1) alone or statement (2) alone are sufficient to answer the question

D. If data given in both (1) & (2) together are not sufficient to answer the question

E. If data in both statements (1) & (2) together are necessary to answer the question

20. On which day the flat was purchased by Rohan in 1996?

(1) Certainly before 18th December, 1996 but definitely not before 15th December, 1996

(2) Certainly after 16th December, 1996 but not later than 19th December, 1996

A. if the data in statement (1) alone are sufficient to answer the question

B. if the data in statement (2) alone are sufficient answer the question

C. if the data either in (1) or (2) alone are sufficient to answer the question

D. if the data even in both the statements together are not sufficient to answer the question

E. if the data in both the statements together are needed

21. Which day of the last week did Satish meet Kapil, at Kapil’s residence?

(1) Kapil was out of town from Monday to Wednesday. He returned on Thursday morning.

(2) On Friday night Satish telephoned his friend to inform that only yesterday he had got approval of Kapil after personally explaining to him all the details.

A. if the data in statement (1) alone are sufficient to answer the question

B. if the data in statement (2) alone are sufficient answer the question

C. if the data either in (1) or (2) alone are sufficient to answer the question

D. if the data even in both the statements together are not sufficient to answer the question

E. If the data in both the statements together are needed

22. Who among P, Q, T, V and M is exactly in the middle when they are arranged in ascending order of their height?

    (1) V is taller than Q but shorter than M.

    (2) T is taller than Q and M but shorter than P

    A. If the data in statement (1) alone are sufficient

    B. If the data in statement (2) alone are sufficient

    C. If the data either in statement (1) alone or in statement (2) alone are sufficient

    D. If the data given in both the statements (1) and (2) together are not sufficient

    E. If the data in both the statements I and II together are necessary

IV. Numerical Reasoning (5 questions)

Directions :

Each question is a sequence of numbers with one or two numbers missing. You have to figure out the logical order of the sequence to find out the missing number(s).

23. 1/8，1/15，1/24，1/35，( )

    A. 1/45

    B. 1/48

    C. 0

    D. 1/76

24. 1，7，8，57，( )

    A. 457

    B. 114

    C. 58

    D. 116

25. 5，16，50，153，( )

    A. 256

    B. 369

    C. 454

D. 463

26. 582，554，526，498，470，( )

A. 442

B. 452

C. 432

D. 462

27. 0，2，2，5，4，7，( )

A. 6

B. 5

C. 4

D. 3

V. Interpretation of Tables and Graphs (8 questions)

Directions :
This is a test on reading and interpretation of data presented in tables and graphs.

Study the following table and answer the questions based on it.

Chart 1

Number of Candidates Appeared, Qualified and Selected in a Competitive Examination from Five States Delhi, H.P, U.P, Punjab and Haryana Over the Years 1994 to 1998

Year	Delhi			H.P			U.P			Punjab			Haryana		
	App	Qual	Sel	App	Qual	Sel	App	Qual	Sel	App	Qual	Sel	App	Qual	Sel
1997	8000	850	94	7800	810	82	7500	720	78	8200	680	85	6400	700	75
1998	4800	500	48	7500	800	65	5600	620	85	6800	600	70	7100	650	75
1999	7500	640	82	7400	560	70	4800	400	48	6500	525	65	5200	350	55
2000	9500	850	90	8800	920	86	7000	650	70	7800	720	84	6400	540	60
2001	9000	800	70	7200	850	75	8500	950	80	5700	485	60	4500	600	75

28. For which state the average number of candidates selected over the years is the maximum?

    A. Delhi

    B. H.P

    C. U.P.

    D. Punjab

29. The percentage of candidates qualified from Punjab over those appeared from Punjab is highestin the year?

A. 1997

B. 1998

C. 1999

D. 2000

30. In the year 1997, which state had the lowest percentage of candidates selected over the candidates appeared?

A. Delhi

B. H.P.

C. U.P.

D. Punjab

31. The number of candidates selected from Haryana during the period under review is approximately what percent of the number selected from Delhi during this period?

A. 79.5%

B. 81%

C. 84.5%

D. 88.5%

Chart 2

A cosmetic company provides five different products. The sales of these five products (in lakh number of packs. during 1995 and 2000 are shown in the following bar graph.

Sales (in lakh number of packs. of five different products of Cosmetic Company during 1995 and 2000

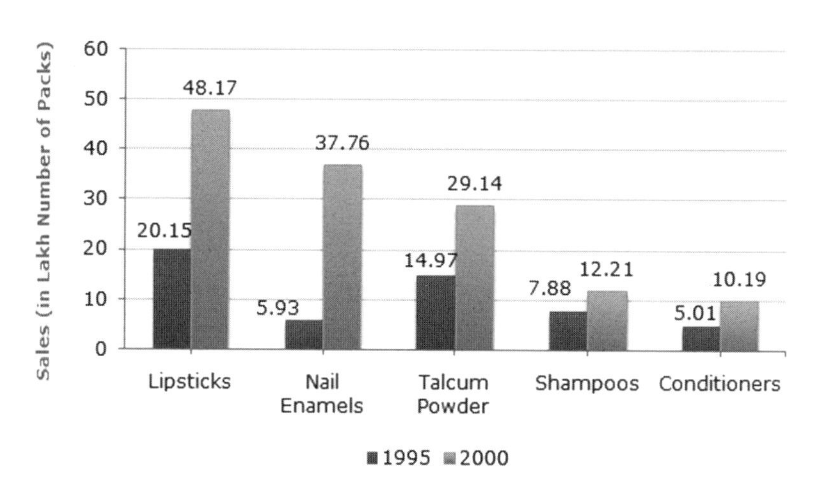

32. The sales of lipsticks in 2000 was by what percent more than the sales of nail enamels in 2000? (rounded off to nearest integer)

    A. 33%

    B. 31%

    C. 28%

    D. 22%

33. During the period 1995-2000, the minimum rate of increase in sales is in the case of?

A. Shampoos

B. Nail enamels

C. Talcum powders

D. Lipsticks

34. What is the approximate ratio of the sales of nail enamels in 2000 to the sales of Talcum powders in 1995?

A. 7:2

B. 5:2

C. 4:3

D. 2:1

35. The sales have increase by nearly 55% from 1995 to 2000 in the case of?

A. Lipsticks

B. Nail enamels

C. Talcum powders

D. Shampoos

全卷完

CRE-APT

文化會社出版社 **CULTURE CROSS LIMITED**

答題紙 ANSWER SHEET

請在此貼上電腦條碼
Please stick the barcode label here

(1) 考生編號 Candidate No.

(2) 考生姓名 Name of Candidate

(3) 考生簽署 Signature of Candidate

宜用H.B.鉛筆作答
You are advised to use H.B. Pencils

考生須依照下圖所示填畫答案：

23 A B C D E

錯填答案可使用潔淨膠擦將筆痕徹底擦去。

切勿摺皺此答題紙

Mark your answer as follows:

23 A B C D E

Wrong marks should be completely erased with a clean rubber.

DO NOT FOLD THIS SHEET

1	A	B	C	D	E	21	A	B	C	D	E
2	A	B	C	D	E	22	A	B	C	D	E
3	A	B	C	D	E	23	A	B	C	D	E
4	A	B	C	D	E	24	A	B	C	D	E
5	A	B	C	D	E	25	A	B	C	D	E
6	A	B	C	D	E	26	A	B	C	D	E
7	A	B	C	D	E	27	A	B	C	D	E
8	A	B	C	D	E	28	A	B	C	D	E
9	A	B	C	D	E	29	A	B	C	D	E
10	A	B	C	D	E	30	A	B	C	D	E
11	A	B	C	D	E	31	A	B	C	D	E
12	A	B	C	D	E	32	A	B	C	D	E
13	A	B	C	D	E	33	A	B	C	D	E
14	A	B	C	D	E	34	A	B	C	D	E
15	A	B	C	D	E	35	A	B	C	D	E
16	A	B	C	D	E	36	A	B	C	D	E
17	A	B	C	D	E	37	A	B	C	D	E
18	A	B	C	D	E	38	A	B	C	D	E
19	A	B	C	D	E	39	A	B	C	D	E
20	A	B	C	D	E	40	A	B	C	D	E

文化會社出版社
投考公務員 模擬試題王

能力傾向試測試
模擬試卷（九）

時間：四十五分鐘

考生須知：

(一) 細讀答題紙上的指示。宣布開考後，考生須首先於適當位置貼上電腦條碼及填上各項所需資料。宣布停筆後，考生不會獲得額外時間貼上電腦條碼。

(二) 試場主任宣布開卷後，考生請檢查試題冊及確定試題冊內的試題。最後會有「**全卷完**」的字眼。

(三) 本試卷各題佔分相等。

(四) **本試卷全部試題均須回答**。為便於修正答案，考生宜用HB鉛筆把答案填畫在答題紙上。錯誤答案可用潔淨膠擦將筆痕徹底擦去。考生須清楚填畫答案，否則會因答案未能被辨認而失分。

(五) 每題只可填畫**一個**答案。如填劃超過一個答案，該題將**不獲評分**。

(六) 答案錯誤，不另扣分。

(七) 未經許多，請勿打開試題冊。

CC-CRE-APT

I. 演繹推理（8題）

請根據以下短文的內容，選出一個或一組推論。請假定短文的內容都是正確的。

1. **表面上看，1982年大學畢業生與1964年的大學畢業生相像，他們都相當保守：衣著良好並對傳統感興趣，尊重他們的父母。但有一個根深蒂固的區別：大多數被調查的1982年的大學畢業生在大學一年級時，宣稱獲得好的收入是他們決定上大學的一個重要原因。**

   上段文字最支持的結論是：

   A. 1964年的大學畢業生實際上比1982年的大學畢業生保守

   B. 多數1964年大學畢業生在大學一、二年級改變了他們上大學的目的

   C. 少於一半的1964年大學畢業生在大學一年級時，宣稱上大學是為了增加他們的收入

   D. 1964年的大學畢業生與1982年的大學畢業生相比較而言對於財政的關心是表面的

2. **婚姻使人變胖。作為此結果證據的是一項調查結果：在13年的婚姻生活中，女性平均胖了43磅，男性平均胖了38磅。為了做進一步的研究，支持這一觀點，下列四個問題中應先做哪一個？**

   A. 「為什麼調查的時間是13年，而不是12或者14年？」

   B. 「調查中的女性和調查的男性態度一樣積極嗎？」

C. 「與調查中年齡相當的單身漢在13年中的體重增加或減少了多少？」

D. 「調查中獲得的體重將維持一生嗎？」

3. **20世紀70年代出現了大學畢業生的過度供給，過度的供給使大學畢業生的平均年收入降到比只持有高中文憑的工人僅高18%的水平。到了20世紀80年代，大學畢業生的平均年收入比只持有高中文憑的工人高43%，儘管20世紀70年代到80年代後期大學畢業生的供給量沒有下降。**

下面哪項，如果在20世紀80年代後期是正確的，最好地調節了上述明顯的分歧？

A. 經濟放慢了，從而使對大學生的需求減少了

B. 高中教育的質量提高了

C. 與20世紀70年代相比，更多的高中為它們提供了職業指導計劃

D. 20年來第一次出現了僅有高中文憑的求職者的過度供給

4. **甲、乙、丙、丁、戊五個學生參加考試，他們的成績之間有關係是：丙沒有乙高，戊沒有丁高，甲高於乙，而丁不如丙高，則成績最高的是：**

A. 甲

B. 乙

C. 丙

D. 丁

5. 雖然菠菜中含有豐富的鈣質，但同時含有大量的漿草酸，漿草酸會有力地阻止人體對於鈣質的吸收。因此，一個人要想攝入足夠的鈣質，就必須用其他含鈣豐富的食物來取代菠菜，至少和菠菜一起食用。
以下哪項如果為真，最能削弱題幹的論證？

A. 大米中不含鈣質；但含有中和漿草酸並改變其性能的鹼性物質

B. 奶制品中的鈣含量要高於菠菜，許多經常食用菠菜的人也同時食用奶製品

C. 在烹飪的過程中，菠菜中受到破壞的漿草酸要略多於鈣

D. 在人的日常飲食中，除了菠菜以外，事實上大量的蔬菜都含有鈣

6. 統計數據正確地揭示：整個20世紀，全球範圍內火山爆發的次數逐年緩慢上升，只有在兩次世界大戰期間，火山爆發的次數明顯下降。科學家同樣正確地揭示：整個20世紀全球火山的活動性處於一個幾乎不變的水平上，這和19世紀的情況形成了鮮明的對比。
如果上述斷定是真的，則以下哪項一定是真的？
(1)如果本世紀不發生兩次世界大戰，全球範圍內火山爆發的次數將無例外地呈逐年緩慢上升的趨勢
(2)火山自身的活動性，並不是造成火山爆發的唯一原因
(3)19世紀全球火山爆發比20世紀頻繁

A. 只有(1)

B. 只有(2)

C. 只有(3)

D. 只有(1)和(2)

7. 售價2元一市斤的洗潔精分為兩種：一種加除臭劑，另一種沒有除臭劑。儘管兩種洗潔精效果相同，但沒有加除臭劑的洗潔精在持久時間方面明顯不如有除臭劑的洗潔精。因為後者：

A. 味道更好些

B. 具有添加劑

C. 從長遠來看更便宜

D. 比其他公司的產品效果好

8. 在一次國際學術會議上，來自四個國家的五位代表被安排坐一張圓桌。為了使他們能夠自由交談，事先了解到情況如下。

(1)甲是中國人，還會說英語

(2)乙是法國人，還會說日語

(3)丙是英國人，還會說法語

(4)丁是日本人，還會說漢語

(5)戊是法國人，還會說德語

請問如何安排?

A. 甲、丙、戊、乙、丁

B. 甲、丁、丙、乙、戊

C. 甲、乙、丙、丁、戊

D. 甲、丙、丁、戊、乙

II. Verbal Reasoning (English) (6 questions)

Directions :

In this test, each passage is followed by three statements (the questions). You have to assume what is stated in the passage is true and decide whether the statements are either:

True (Box A): the statement is already made or implied in the passage, or follows logically from the passage.

False (Box B): the statement contradicts what is said, implied by, or follows logically from the passage.

Can' t tell (Box C): there is insufficient information in the passage to establish whether the statement is true or false.

Passage 1 (Question 9 to 11):

Crude oil (also known as petroleum) is a type of fossil fuel found beneath the earth's surface. It is formed by the gradual build-up of fossilised organic materials such as algae and plankton. As more layers build up, the bottom most layers are heated and subject to pressure, with the combination of heat and pressure leading to the matter transforming into the waxy substance kerogen. Following even more prolonged exposure to heat and pressure, the kerogen eventually becomes transformed into liquid and gases via the catagenesis process. The formation of crude oil occurs from this pyrolysis (heating) process. The range of heat at which kerogen becomes crude oil is called the oil window. Below this range the crude oil remains kerogen and above this point the crude oil becomes a natural gas.

9. If the temperature is too low, crude oil remains in a solid state, whereas if it is too hot, it becomes a gas.

10. Crude oil is non-renewable.

11. Kerogen becomes crude oil after further heating and pressurisation in the catagenesis process.

Passage 2 (Question 12 to 14):

Wine is an alcoholic beverage made from fermented grapes. Wine is often associated most closely with France and with some justification. France is easily the biggest producer of wine, biggest consumer (on a per capita basis) and usually the biggest exporter of it. The process of creating wine is called vinification or winemaking. Grapes are crushed and then fermented using yeast. Yeast has the effect of consuming the sugars in grapes and then producing alcohol. Carbon dioxide is also produced as part of this process, but it is normally not captured. There are additional processes depending on the type of wine that is to be produced. For example, red wine undergoes a secondary fermentation which includes the conversion of malic acid into lactic acid, which aims to soften the taste of the wine. Red wine may also be transferred to oak barrels for maturity to induce an 'oakiness' on the produce.

12. Carbon dioxide, which is a by-product of the winemaking process, is sometimes captured and then used to carbonate other beverages.

13. White wine may be transferred to oak barrels to create an 'oakiness' flavor dimension.

14. Vinification is the same thing as winemaking.

III. Data Sufficiency Test (8 questions)

Directions : In this test, you are required to choose a combination of clues to solve a problem.

15. How much was the total sale of the company?

    (1) The company sold 8000 units of product A each costing Rs.25

    (2) This company has no other product line

    A. if the data in statement (1) alone are sufficient to answer the question

    B. if the data in statement (2) alone are sufficient answer the question

C. if the data either in (1) or (2) alone are sufficient to answer the question

D. if the data even in both the statements together are not sufficient to answer the question

E. If the data in both the statements together are needed

16. Who is the immediate right of P among five persons P, Q, R, S and T facing North?

(1) R is third to the left of Q; P is second to the right of R

(2) Q is to the immediate left of T, who is second to the right of P

A. If the data in statement (1) alone are sufficient

B. If the data in statement (2) alone are sufficient

C. If the data either in statement (1) alone or in statement (2) alone are sufficient

D. If the data given in both the statements (1) and (2) together are not sufficient

E. If the data in both the statements (1) and (2) together are necessary

17. How is B related to H?

(1) H' s mother is sister of B' s father' s son

(2) B is father of T

A. If the data in statement (1) alone are sufficient

B. If the data in statement (2) alone are sufficient

C. If the data either in statement (1) alone or in statement (2) alone are sufficient

D. If the data given in both the statements (1) and (2) together are not sufficient

E. If the data in both the statements (1) and (2) together are necessary

18. How is C related to A?

(1) A is the son of B, D is the husband of B

(2) F is married to C, C has 2 children D and E.D is the father of A

A. If data in the statement (1) alone is sufficient to answer the question.

B. If data in the statement (2) alone is sufficient to answer the question.

C. If data either in the statement (1) alone or statement (2) alone are sufficient to answer the question.

D. If data given in both (1) & (2) together are not sufficient to answer the question.

E. If data in both statements (1) & (2) together are necessary to answer the question.

19. A, B, C, D and E are sitting in a row. B is between A and E. Who among them is in the middle?

(1) A is left of B and right of D

(2) C is at the right end

A. if the data in statement (1) alone are sufficient to answer the question

B. if the data in statement (2) alone are sufficient answer the question

C. if the data either in (1) or (2) alone are sufficient to answer the question

D. if the data even in both the statements together are not sufficient to answer the question

E. If the data in both the statements together are needed

20. At present, how many villagers are voters in village 'X' ?

(1) There were 860 voters in village 'X' in the list prepared for the last election.

(2) The present list of village 'X' has 15% more voters than the list for the last election.

A. If the data in statement (1) alone are sufficient

B. If the data in statement (2) alone are sufficient

C. If the data either in statement (1) alone or in statement (2) alone are sufficient

D. If the data given in both the statements (1) and (2) together are not sufficient

E. If the data in both the statements (1) and (2) together are necessary

21. What is the meaning of “nic” in a certain code language.

(1) In that code language “pat nic no ran” means “what is your name”

(2) In that code language “nic sa ranja” means “my name is Shambhu”

A. If the data in statement (1) alone are sufficient

B. If the data in statement (2) alone are sufficient

C. If the data either in statement (1) alone or in statement (2) alone are sufficient

D. If the data given in both the statements (1) and (2) together are not sufficient

E. If the data in both the statements (1) and (2) together are necessary

22. How is ‘No’ coded the code language?

(1) ‘Ne Pa Sic Lo’ means ‘But No None And’ and ‘Pa Lo Le Ne’ means ‘If None And But’

(2) ‘Le Se Ne Sic’ means ‘If No None Will’ and ‘Le Pi Se Be’ means ‘Not None If All’

A. If the data in statement (1) alone are sufficient

B. If the data in statement (2) alone are sufficient

C. If the data either in statement (1) alone or in statement (2) alone are sufficient

D. If the data given in both the statements (1) and (2) together are not sufficient

E. If the data in both the statements (1) and (2) together are necessary

IV. Numerical Reasoning (5 questions)

Directions :
Each question is a sequence of numbers with one or two numbers missing. You have to figure out the logical order of the sequence to find out the missing number(s).

23. 44，52，68，76，92，( )

A. 104

B. 116

C. 124

D. 128

24. 3，4，8，17，( )，58

A. 16

B. 26

C. 33

D. 45

25. 64，48，36，27，81/4，( )

A. 97/6

B. 123/38

C. 179/12

D. 243/16

26. 1，-4，4，8，40，(　)

A. 160

B. 240

C. 320

D. 480

27. 2，2，7，9，16，20，(　)

A. 28

B. 29

C. 30

D. 31

V. Interpretation of Tables and Graphs (8 questions)

Directions :

This is a test on reading and interpretation of data presented in tables and graphs.

Chart 1

The circle-graph given here shows the spendings of a country on various sports during a particular year. Study the graph carefully and answer the questions given below it.

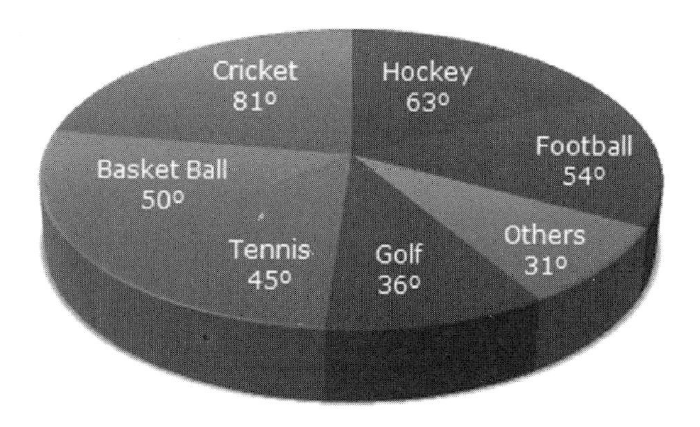

28. What percent of total spending is spent on Tennis?

A. 12(½)%

B. 22(½)%

C. 25%

D. 45%

29. How much percent more is spent on Hockey than that on Golf?

A. 27%

B. 35%

C. 37.5%

D. 75%

30. If the total amount spent on sports during the year be Rs. 1,80,00,000 , the amount spent on Basketball exceeds on Tennis by:

A. Rs. 2,50,000

B. Rs. 3,60,000

C. Rs. 3,75,000

D. Rs. 4,10,000

31. If the total amount spent on sports during the year was Rs. 2 crores, the amount spent on Cricket and Hockey together was:

A. Rs. 800,000

B. Rs. 8,000,000

C. Rs. 12,000,000

D. Rs. 16,000,000

Chart 2

Answer the questions based on the given line graph.

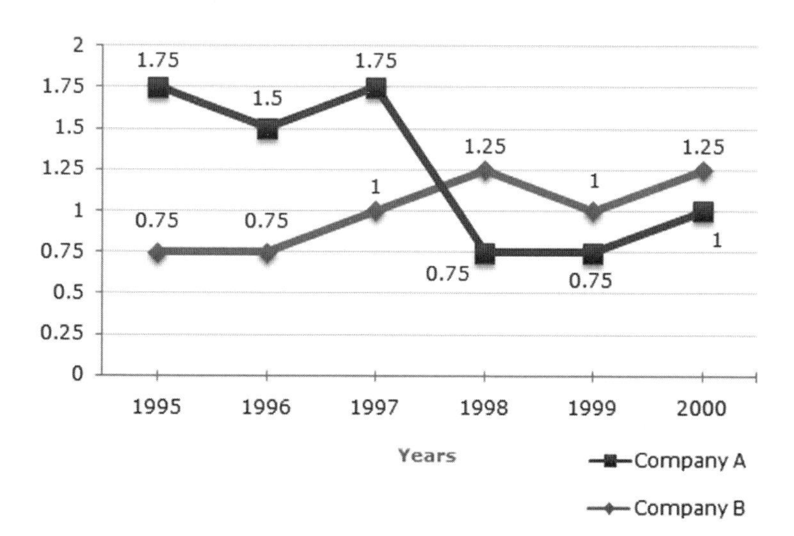

Ratio of Exports to Imports (in terms of money in Rs. crores) of Two Companies Over the Years

32. In how many of the given years were the exports more than the imports for Company A?

    A. 2

    B. 3

    C. 4

    D. 5

33. If the exports of Company A in 1998 were Rs. 237 crores, what was the amount of imports in that year?

A. Rs. 189.6 crores

B. Rs. 243 crores

C. Rs. 281 crores

D. Rs. 316 crores

34. In 1995, the export of Company A was double that of Company B. If the imports of Company A during the year was Rs. 180 crores, what was the approximate amount of imports pf Company B during that year?

A. Rs. 190 crores

B. Rs. 210 crores

C. Rs. 225 crores

D. Cannot be determined

35. In which year(s) was the difference between impors and exports of Company B the maximum?

A. 2000

B. 1996

C. 1998 and 2000

D. Cannot be determined

全卷完

PART TWO
《能力傾向試》模擬試卷答案

模擬試卷（一）答案

1. D
2. A
3. D
4. B
5. C
6. D
7. D
8. D
9. C
10. B
11. C
12. C
13. A
14. C
15. A
16. D
17. D
18. C
19. E
20. C
21. E
22. E
23. A
24. B
25. B
26. C
27. D
28. C
29. B
30. C
31. B
32. D
33. D
34. D
35. C

模擬試卷（二）答案

1. A
2. D
3. C
4. C
5. A
6. B
7. C
8. D
9. C
10. C
11. B
12. B
13. C
14. A
15. E
16. D
17. C
18. B
19. C
20. D
21. D

22. E
23. D
24. C
25. B
26. B
27. C
28. D
29. C
30. D
31. D
32. A
33. C
34. B
35. C

模擬試卷（三）答案

1. B
2. B
3. B
4. C
5. B
6. B
7. A
8. A
9. A
10. A
11. B
12. C
13. B
14. A
15. D
16. E
17. E
18. E
19. A
20. D
21. E
22. C
23. A
24. A
25. C
26. B
27. A
28. D
29. D
30. B
31. D
32. B
33. D
34. C
35. D

模擬試卷（四）答案

1. D
2. B
3. A
4. A
5. B

6. C
7. D
8. D
9. A
10. C
11. A
12. B
13. A
14. A
15. D
16. B
17. D
18. D
19. E
20. B
21. C
22. D
23. C
24. C
25. B
26. B
27. A
28. C
29. D
30. D
31. A
32. A
33. B
34. A
35. D

模擬試卷（五）答案

1. C
2. C
3. C
4. B
5. C
6. C
7. B
8. B
9. C
10. A
11. B
12. C
13. B
14. B
15. B
16. C
17. D
18. E
19. D
20. D
21. B
22. E
23. D
24. D
25. C
26. A
27. D
28. B
29. B

30. C
31. C
32. D
33. B
34. B
35. D

模擬試卷（六）答案

1. D
2. A
3. D
4. B
5. A
6. C
7. C
8. A
9. B
10. A
11. A
12. A
13. C
14. A
15. A
16. D
17. C
18. E
19. C
20. A
21. D
22. B
23. A
24. B
25. A
26. C
27. B
28. D
29. B
30. C
31. D
32. C
33. D
34. C
35. D

模擬試卷（七）答案

1. A
2. B
3. A
4. C
5. D
6. A
7. C
8. C
9. A
10. B
11. C
12. A
13. B

14. C
15. E
16. E
17. E
18. E
19. C
20. E
21. E
22. E
23. C
24. C
25. D
26. B
27. C
28. C
29. D
30. D
31. B
32. C
33. B
34. B
35. C

模擬試卷（八）答案

1. B
2. C
3. D
4. B
5. C
6. D
7. A
8. C
9. B
10. C
11. C
12. B
13. A
14. C
15. A
16. B
17. A
18. A
19. E
20. E
21. B
22. E
23. B
24. A
25. D
26. A
27. A
28. A
29. D
30. D
31. D
32. C
33. A
34. B
35. D

模擬試卷（九）答案

1. C
2. C
3. D
4. A
5. A
6. B
7. A
8. A
9. A
10. C
11. A
12. C
13. C
14. B
15. E
16. C
17. E
18. D
19. E
20. E
21. D
22. A
23. B
24. C
25. D
26. C
27. B
28. A
29. D
30. A
31. B
32. B
33. D
34. B
35. D

PART THREE

有用資料

公務員職系要求一覽

	職系	入職職級	英文運用	中文運用	能力傾向測試
1.	會計主任	二級會計主任	二級	二級	及格
2.	政務主任	政務主任	二級	二級	及格
3.	農業主任	助理農業主任／農業主任	一級	一級	及格
4.	系統分析／程序編製主任	二級系統分析／程序編製主任	二級	二級	及格
5.	建築師	助理建築師／建築師	一級	一級	及格
6.	政府檔案處主任	政府檔案處助理主任	二級	二級	-
7.	評税主任	助理評税主任	二級	二級	及格
8.	審計師	審計師	二級	二級	及格
9.	屋宇裝備工程師	助理屋宇裝備工程師／屋宇裝備工程師	一級	一級	及格
10.	屋宇測量師	助理屋宇測量師／屋宇測量師	一級	一級	及格
11.	製圖師	助理製圖師／製圖師	一級	一級	-
12.	化驗師	化驗師	一級	一級	及格
13.	臨床心理學家（衞生署、入境事務處）	臨床心理學家（衞生署、入境事務處）	一級	一級	-
14.	臨床心理學家（懲教署、香港警務處）	臨床心理學家（懲教署、香港警務處）	二級	二級	-
15.	臨床心理學家（社會福利署）	臨床心理學家（社會福利署）	二級	二級	及格
16.	法庭傳譯主任	法庭二級傳譯主任	二級	二級	及格
17.	館長	二級助理館長	二級	二級	-
18.	牙科醫生	牙科醫生	一級	一級	-
19.	營養科主任	營養科主任	一級	一級	-
20.	經濟主任	經濟主任	二級	二級	-
21.	教育主任（懲教署）	助理教育主任（懲教署）	一級	一級	-

	職系	入職職級	英文運用	中文運用	能力傾向測試
22.	教育主任（教育局、社會福利署）	助理教育主任（教育局、社會福利署）	二級	二級	-
23.	教育主任（行政）	助理教育主任（行政）	二級	二級	-
24.	機電工程師（機電工程署）	助理機電工程師／機電工程師（機電工程署）	一級	一級	及格
25.	機電工程師（創新科技署）	助理機電工程師／機電工程師（創新科技署）	一級	一級	-
26.	電機工程師（水務署）	助理機電工程師／機電工程師（水務署）	一級	一級	及格
27.	電子工程師（民航署、機電工程署）	助理電子工程師／電子工程師（民航署、機電工程署）	一級	一級	及格
28.	電子工程師（創新科技署）	助理電子工程師／電子工程師（創新科技署）	一級	一級	-
29.	工程師	助理工程師／工程師	一級	一級	及格
30.	娛樂事務管理主任	娛樂事務管理主任	二級	二級	及格
31.	環境保護主任	助理環境保護主任／環境保護主任	二級	二級	及格
32.	產業測量師	助理產業測量師／產業測量師	一級	一級	-
33.	審查主任	審查主任	二級	二級	及格
34.	行政主任	二級行政主任	二級	二級	及格
35.	學術主任	學術主任	一級	一級	-
36.	漁業主任	助理漁業主任／漁業主任	一級	一級	及格
37.	警察福利主任	警察助理福利主任	二級	二級	-
38.	林務主任	助理林務主任／林務主任	一級	一級	及格
39.	土力工程師	助理土力工程師／土力工程師	一級	一級	及格
40.	政府律師	政府律師	二級	一級	-
41.	政府車輛事務經理	政府車輛事務經理	一級	一級	-

	職系	入職職級	英文運用	中文運用	能力傾向測試
42.	院務主任	二級院務主任	二級	二級	及格
43.	新聞主任（美術設計）／（攝影）	助理新聞主任（美術設計）／（攝影）	一級	一級	-
44.	新聞主任（一般工作）	助理新聞主任（一般工作）	二級	二級	及格
45.	破產管理主任	二級破產管理主任	二級	二級	及格
46.	督學（學位）	助理督學（學位）	二級	二級	-
47.	知識產權審查主任	二級知識產權審查主任	二級	二級	及格
48.	投資促進主任	投資促進主任	二級	二級	-
49.	勞工事務主任	二級助理勞工事務主任	二級	二級	及格
50.	土地測量師	助理土地測量師／土地測量師	一級	一級	-
51.	園境師	助理園境師／園境師	一級	一級	及格
52.	法律翻譯主任	法律翻譯主任	二級	二級	-
53.	法律援助律師	法律援助律師	二級	一級	及格
54.	圖書館館長	圖書館助理館長	二級	二級	及格
55.	屋宇保養測量師	助理屋宇保養測量師／屋宇保養測量師	一級	一級	及格
56.	管理參議主任	二級管理參議主任	二級	二級	及格
57.	文化工作經理	文化工作副經理	二級	二級	及格
58.	機械工程師	助理機械工程師／機械工程師	一級	一級	及格
59.	醫生	醫生	一級	一級	-
60.	職業環境衞生師	助理職業環境衞生師／職業環境衞生師	二級	二級	及格
61.	法定語文主任	二級法定語文主任	二級	二級	-
62.	民航事務主任（民航行政管理）	助理民航事務主任（民航行政管理）／民航事務主任（民航行政管理）	二級	二級	及格
63.	防治蟲鼠主任	助理防治蟲鼠主任／防治蟲鼠主任	一級	一級	及格

	職系	入職職級	英文運用	中文運用	能力傾向測試
64.	藥劑師	藥劑師	一級	一級	-
65.	物理學家	物理學家	一級	一級	及格
66.	規劃師	助理規劃師／規劃師	二級	二級	及格
67.	小學學位教師	助理小學學位教師	二級	二級	-
68.	工料測量師	助理工料測量師／工料測量師	一級	一級	及格
69.	規管事務經理	規管事務經理	一級	一級	-
70.	科學主任	科學主任	一級	一級	-
71.	科學主任（醫務）（衞生署）	科學主任（醫務）（衞生署）	一級	一級	-
72.	科學主任（醫務）（食物環境衞生署）	科學主任（醫務）（食物環境衞生署）	一級	一級	及格
73.	管理值班工程師	管理值班工程師	一級	一級	-
74.	船舶安全主任	船舶安全主任	一級	一級	-
75.	即時傳譯主任	即時傳譯主任	二級	二級	-
76.	社會工作主任	助理社會工作主任	二級	二級	及格
77.	律師	律師	二級	一級	-
78.	專責教育主任	二級專責教育主任	二級	二級	-
79.	言語治療主任	言語治療主任	一級	一級	-
80.	統計師	統計師	二級	二級	及格
81.	結構工程師	助理結構工程師／結構工程師	一級	一級	及格
82.	電訊工程師（香港警務處）	助理電訊工程師／電訊工程師（香港警務處）	一級	一級	-
83.	電訊工程師（通訊事務管理局辦公室）	助理電訊工程師／電訊工程師（通訊事務管理局辦公室）	一級	一級	及格
84.	電訊工程師（香港電台）	高級電訊工程師／助理電訊工程師／電訊工程師（香港電台）	一級	一級	-

	職系	入職職級	英文運用	中文運用	能力傾向測試
85.	電訊工程師（消防處）	高級電訊工程師（消防處）	一級	一級	-
86.	城市規劃師	助理城市規劃師／城市規劃師	二級	二級	及格
87.	貿易主任	二級助理貿易主任	二級	二級	及格
88.	訓練主任	二級訓練主任	二級	二級	及格
89.	運輸主任	二級運輸主任	二級	二級	及格
90.	庫務會計師	庫務會計師	二級	二級	及格
91.	物業估價測量師	助理物業估價測量師／物業估價測量師	一級	一級	及格
92.	水務化驗師	水務化驗師	一級	一級	及格

看得喜 放不低

創出喜閱新思維

書名 投考公務員 能力傾向試模擬試卷精讀 Aptitude Test: Mock Paper
ISBN 978-988-74807-1-6
定價 HK$128
出版日期 2021年3月
作者 Fong Sir
責任編輯 Y.T.
版面設計 梁文俊
出版 文化會社有限公司
電郵 editor@culturecross.com
網址 www.culturecross.com
發行 香港聯合書刊物流有限公司
地址：香港新界大埔汀麗路36號中華商務印刷大廈3樓
電話：（852）2150 2100
傳真：（852）2407 3062